Estimator

The *Estimator's Pocket Book, Second Edition* is a concise and practical reference covering the main pricing approaches, as well as useful information such as how to process sub-contractor quotations, tender settlement and adjudication. It is fully up to date with NRM2 throughout, features a look ahead to NRM3 and describes the implications of BIM for estimators.

It includes instructions on how to handle:

- the NRM order of cost estimate;
- unit-rate pricing for different trades;
- pro-rata pricing and dayworks;
- builders' quantities;
- approximate quantities.

Worked examples show how each of these techniques should be carried out in clear, easy-to-follow steps. This is the indispensable estimating reference for all quantity surveyors, cost managers, project managers and anybody else with estimating responsibilities. Particular attention is given to NRM2, but the overall focus is on the core estimating skills needed in practice. Updates to this edition include a greater reference to BIM, an update on the current state of the construction industry as well as up-to-date wage rates, legislative changes and guidance notes.

Duncan Cartlidge is a chartered surveyor and construction procurement consultant with extensive experience in the delivery and management of built assets, as well as providing education and training to a wide range of built-environment professionals.

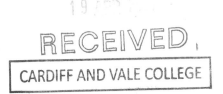

Estimator's Pocket Book

Second Edition

Duncan Cartlidge FRICS

Routledge
Taylor & Francis Group

LONDON AND NEW YORK

Second edition published 2019
by Routledge
2 Park Square, Milton Park, Abingdon, Oxon OX14 4RN

and by Routledge
52 Vanderbilt Avenue, New York, NY 10017

Routledge is an imprint of the Taylor & Francis Group, an informa business

First edition published by Routledge 2013

British Library Cataloguing-in-Publication Data
A catalogue record for this book is available from the British Library

Library of Congress Cataloging-in-Publication Data
Names: Cartlidge, Duncan P., author.
Title: Estimator's pocket book / Duncan Cartlidge.
Description: Second edition. | Abingdon, Oxon ; New York, NY :
    Routledge, 2019.
Identifiers: LCCN 2018038405| ISBN 9781138366695 (hardback) |
    ISBN 9781138366701 (paperback) | ISBN 9780429430176 (ebook)
Subjects: LCSH: Building—Estimates—Handbooks, manuals, etc.
Classification: LCC TH435 .C363 2019 | DDC 692/.5—dc23
LC record available at https://lccn.loc.gov/2018038405

ISBN: 978–1–138–36669–5 (hbk)
ISBN: 978–1–138–36670–1 (pbk)
ISBN: 978–0–429–43017–6 (ebk)

Typeset in Goudy
by Swales & Willis Ltd, Exeter, Devon, UK

Dedication
Harry and Darcy

Contents

Preface

Estimating: precision guesswork based on unreliable data provided by those with questionable knowledge.

Estimating is one of the core skills of construction professionals, and success as an estimator relies, among other things, on a sound knowledge of the relevant technologies. However, in recent years when main contractors have become, on the whole, managing contractors relying heavily, if not totally, on sub-contractors to price work sections, estimating from basic principles, like measurement, has tended to be left out of university courses and training programmes. Not only are estimating skills required for preparing bids, but also during the post-contract phase they are essential when negotiating rates and prices for settling final accounts and claims. When times are hard and work scarce, then the estimator's role becomes even more crucial as they are the people within a contracting organisation who win work and keep the cash flow flowing. In 2018 major UK contractors declared profit margins of less than 1% and with such little wiggle room it is even more important for estimating to be accurate and a true reflection of the cost of the works. Estimators are to be found in a wide range of organisations across the building, civil engineering, and mechanical and electrical sectors, working for main contractors, sub-contractors and component manufacturers.

Estimating tends to be a specialism that people come to during their professional lives and learn on the job. To the author's knowledge, there aren't any degree programmes or courses in estimating in the UK, hence the need for a pocket book for constructional professionals and students.

The increased pervasiveness of systems such as Building Information Management (BIM) claim to reduce the need for the manual measurement, quantification and costing of building projects. Whether this is the future is anyone's guess; however, the time when sophisticated software will ever be able to carry out tender adjudication, assess risk and negotiate with the bank is perhaps a few years away yet!

Fifty years or so ago, the majority of work was based on a fully detailed bill of quantities, prepared by a quantity surveyor employed by the client; typically the tender would have been based on detailed bills of quantities. In the modern construction market, bids are now based on a variety of approaches – including work packages, cost plans, and drawings and specification – requiring contractors and sub-contractors to be flexible in their approach to formulating a bid and identifying risk.

The examples in this pocket book are based entirely on the RICS *New Rules of Measurement 2: Detailed Measurement for Building Works*.

Duncan Cartlidge
www.duncancartlidge.co.uk

1

Approaches to pricing

Estimators can be found in a wide range of organisations across the building, civil engineering, and mechanical and electrical sectors, working for main contractors, sub-contractors and component manufacturers. Estimating, like measurement, is one of the core skills of quantity surveyors and construction managers, and success as an estimator relies on a sound knowledge of the relevant technologies. However, in the recent past, with the increasing use of sub-contractors, estimating from basic principles has tended to be left out of university courses and training programmes. Estimators carry a great responsibility, as they are the people within a contracting organisation who win work, keeping the cash flow flowing and the profit margins intact. With major UK contractors working on paper-thin margins, it's even more important that estimators get their sums right.

Estimating tends to be a specialism that people come to during their professional lives and learn on the job. To the author's knowledge, there aren't any degree programmes or courses in estimating in the UK. Estimators are responsible for the preparation of tenders or bids that are submitted within a given time frame, and usually in competition with other contractors or sub-contractors, to win work. The format and extent of the information that estimators are supplied with to prepare a bid will vary according to the procurement strategy. Fifty years or so ago, the majority of work was based on a fully detailed bill of quantities, prepared by a quantity surveyor employed by the client. Typically the tender would have been based on:

- two copies of the bills of quantities;
- indicative drawings;
- form of tender;
- instructions for receipt of tender; and
- technical reports/studies.

More recently, the format and the completeness of tender documentation have tended to vary considerably with the traditional bills of quantities being used less and less.

Generally, there are two strategies for obtaining a bid, these are by

- negotiation; or
- competition.

NEGOTIATION

Negotiation involves the client's and contractor's representatives sitting down and negotiating a price for a project without the benefit of competition from other contractors. It is viewed with suspicion by many who consider that, without competition, a contractor will take advantage of the situation and negotiate a higher than market price as the client has no alternative other than to accept it. Anecdotal evidence suggests that negotiation results in bids approximately 11% higher than bids obtained through competition. However, the advantage of negotiation is that the estimating/bidding process can be shorter than with the competitive approach and that, if there is trust between the parties, the tender can be no more costly than by introducing competition. Due to the potential to deliver a project earlier than otherwise would have been the case, project finance may be recouped earlier and finance charges reduced. In this situation, the estimator will be involved in providing the negotiator with data on material, labour and plant costs, etc.

COMPETITION

In the first part of the 21st century, the majority of work in the construction industry is won through competition, with three or four contractors or sub-contractors submitting confidential bids; it's a system that nearly always guarantees that the lowest price wins and ignores wider issues of value for money. The most popular procurement routes that use competition are:

- single-stage competitive tendering;
- two-stage competitive tendering; and
- design and build.

It's true to say that asking for competing contractors to submit bids based on largely incomplete information without the involvement of the supply chain is not the most effective way to arrive at a reliable bid. In particular, the Westminster and Scottish governments have, during the past few years, been highly critical of this approach and have piloted a number of alternatives. In *Construction 2025 Industrial Strategy: Government and Industry in Partnership*, HM Government procurement was described as one of the main barriers to innovation, stating that 'the nature of construction procurement frequently restricts collaboration between client and supply chain'.

In attempt to improve the situation, three new models of procurement were proposed by government, namely:

- cost-led procurement;
- integrated project insurance; and
- two-stage open book.

In an attempt to ensure the price is built up against an outline declared budget and client requirements, the common principles of these models are to:

- engage with the supply chain and embrace early contractor involvement and a high level of supply-chain integration; and
- apply a robust review/risk-analysis process.

It is claimed that a range of pilot projects delivered savings of £840 million.

TRADITIONAL SINGLE-STAGE PROCUREMENT

The approach to the estimating and bidding process will vary according to the procurement strategy adopted by the client for a project. The procurement route

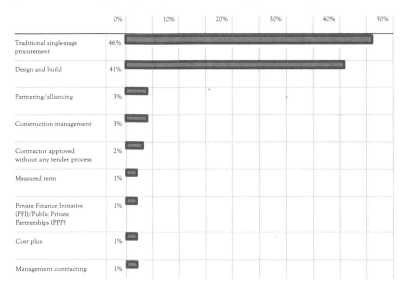

Figure 1.1 Procurement routes

Source: NBS National Construction Contracts and Law Report 2018

will also affect the type of documentation and other tender information received by the estimator. During the 1960s in the UK, the traditional strategy described below was the most commonly used form of construction procurement. It involved a bill of quantities, with approximately 60%-plus of all contracts being let on this basis in both the public and private sectors. In recent times, client pressure has seen its popularity decrease to 46% (NBS National Construction Contracts and Law 2018) (see Figure 1.1). This approach to procurement is also sometimes referred to as 'architect-led' procurement, as traditionally the client has chosen and approached an architect in the first instance, and it is then the architect who assembles the rest of the design team: structural engineer, services engineer, quantity surveyor, etc. The chief characteristics of traditional single-stage competitive tendering are:

- it is based on a linear process with little or no parallel working, resulting in a sometimes lengthy and costly procedure;
- competition or tendering cannot be commenced until the design is completed;
- the tender is based on fully detailed bills of quantities; and
- the lack of contractor involvement in the design process – with the design and technical development being carried out by the client's consultants, unlike some other strategies described later.

Other procurement paths have attempted to shorten the procurement process with the introduction of parallel working between the stages of client brief, design, competition and construction. During stages 2, 3 and 4 – Concept, Developed Design, and Technical Design of the Royal Institution of British Architects (RIBA) Plan of Work – the quantity surveyor should draw up a list of tenderers. The list should comprise three contractors/sub-contractors that are to be approached to carry out the work. The list may be extended to six in the public sector, although it is increasingly difficult to find this number of competent contractors available at the same time. During the preparation of the bills of quantities, the quantity surveyor should contact prospective firms, which have the approval of the client and the architect, to determine whether they are available to bid for the project. In the first instance, this is done by telephoning the chief estimator of a prospective contractor and giving brief details of the proposed project including the approximate value, date for dispatch of documents and the starting date. The decision on whether or not to tender for a project will be influenced by:

- general workload;
- future commitments;
- market conditions;
- capital;
- risk;
- prestige;
- estimating workload; and

- timing – one of the decisions that will have to be taken by the design team is the length of the contract period, a critical calculation, as successful contractors will be liable to pay damages in the event of delays and non-completion.

If the contractor is interested and available for the project, the enquiry is followed up with a letter giving the following information:

- name of client, architect and other lead consultants;
- name and type of project;
- location;
- approximate value;
- brief description;
- date for dispatch and return of tenders;
- start on site and contract duration;
- form of contract; and
- particular conditions applying to the contract.

When the single-stage procurement route is used, the estimating department will receive the following documentation during RIBA Plan of Work Stages 4 and 5, Technical Design, and Construction:

- two copies of the bills of quantities, one bound and one unbound. The bound copy is for pricing and submission, the unbound copy is to allow the contractor to split the bills up into trades so that they can be sent to sub-contractors for pricing;
- indicative drawings on which the bills of quantities were prepared;
- the form of tender – a statement of the tender's bid; and
- instructions (precise time and place) along with an envelope for the return of the tender.

Only indicative drawings are sent out with the tender documents, however details are also given as to where tenderers may inspect a full set of drawings, usually the architect's office. Details should also be included as to times when tenderers may gain access to the site.

The usual time given to contractors for the return of tenders is four weeks, although for particularly complex project this may be six weeks. The tender documents contain precise details for the return of the tender, usually 12 noon on the selected date at the architect's office. Note that the system described above is the one used in England and Wales; in Scotland it is slightly different in so far that tenderers submit a priced bill of quantities, not just a form of tender for consideration.

It is common during the tender period for contractors to raise queries with the quantity surveyor on the tender documentation. If errors, omissions or other anomalies come to light, then the quantity surveyor, once the problem has been resolved, must communicate the result in writing to all tenderers, so that they continue to work on the same basis.

Completed tenders are submitted to the architect's office on or before the required deadline stated in the instructions. Any tenders that are submitted late should be discarded as it is thought it may have been possible to gain an unfair advantage. In England and Wales, a submitted tender may be withdrawn any time prior to acceptance.

The tenders are opened and one selected; but what criteria should be used for selection? Traditionally, the lowest price was chosen on the basis that this provides the client with the best value for money; however, there is an increasing realization that cheapest price may not provide clients with best value to money over the life cycle of a building.

Typical weighted criteria on which tenders are judged could include some or all of the following items:

- green issues;
- technical and professional ability;
- technical resources;
- methodology; and
- price.

No more than five criteria should be used.

When price is the most important of the criteria, one tender is chosen and another bid is selected as a reserve; both selected and reserve tenderers are informed with the first choice bidder being instructed to submit fully priced bills of quantities for checking as soon as possible, usually within 48 hours. The unsuccessful tenderers are told that they have not won the project. Bills of quantities are very comprehensive documents, containing thousands of individual prices and other data relating to a contract, and it is now the responsibility of the quantity surveyor to check the bills and prepare a tender report for the client.

TENDER EVALUATION

The tender evaluation should be treated as a confidential exercise. Once the quantity surveyor is handed the priced bills of quantities, the due diligence and evaluation can begin. The checks that should be carried out are as follows:

- Arithmetical checks confirm that prices have been extended correctly, that page totals are accurate and that page totals have been correctly carried forwarded to trade/element summaries and from there to the main summaries. It is not uncommon for extended bill rates to have errors, and any errors should be noted. In the RIBA Tendering Practice Note 2017, there are two accepted ways to deal with errors in computation:

- ○ The tenderer should be given details of the errors and allowed the opportunity of confirming or withdrawing the offer. In other words, the tenderer would proceed as if the error had not been made. If the tenderer withdraws the offer, the second reserve bills of quantities should be called for and examined. An amendment, signed by both parties to the contract, should be added to the priced bills indicating that all measured rates are to be reduced or increased in the same proportion as the corrected total of priced items exceeds or falls short of such items.
- ○ In the second alternative, the tenderer should be given the opportunity of confirming the offer or of amending it to correct genuine errors. Should the tenderer elect to amend the offer and the revised tender is no longer the lowest, the offer of the reserve tendered should be examined. If the errors are accepted by the tenderer, then the rates in question should be altered and initialled.

- Any items left unpriced should be queried with the tenderer.
- The general level of pricing should be examined and, in particular:

  - ○ Check that similar items that appear in different parts of the bills are priced consistently. For example, excavation items that appear in substructure, external works and drainage.
  - ○ Check that items marked as *provisional* – and therefore to be remeasured during the course of the works – have been priced at rates that are consistent with the rest of the works.
  - ○ Check for caveats inserted by the contractor. For example: 'Removal and disposal of all protective materials has not been included in the prices.'

Once these checks have been carried out and the tender is free of errors, the quantity surveyor can recommend acceptance.

The advantages of single-stage competitive tendering are:

- it is well known and trusted by the industry;
- it ensures competitive fairness;
- for the public sector, it allows audits and accountability to be carried out; and
- it is a valuable post-contract tool that makes the valuation of variations and the preparation of interim payments easier.

The disadvantages are:

- it is a slow, sequential process;
- there is no contractor or supply chain involvement;
- pricing can be manipulated by tenderers; and
- there are high professional fees.

A MODERN VARIATION ON THE TRADITIONAL SINGLE-STAGE SELECTIVE TENDERING THEME

- A fully developed business case is subjected to rigorous cost planning by consultants in conjunction with the client.
- Suitable contractors, say two or three, are pre-selected using appropriate selection criteria, which could be:

  - commitment to supply-chain management;
  - the ability to guarantee life-cycle costs; and
  - capability to deliver the project, etc.

- The pre-selected contractors are asked to fully cost the project proposals and submit their best and final offer (BFO).
- On the basis of the BFO, a contractor is selected and enters into contract guaranteeing the BFO price and, in some cases, whole-life costs for a predetermined period.
- Work starts on site.

TWO-STAGE COMPETITIVE TENDERING

First used widely in the 1970s and based on the traditional single-stage competitive tendering, i.e. bills of quantities and drawings being used to obtain a lump-sum bid. Advantages include early contractor involvement, a fusion of the design/procurement/construction phases, and a degree of parallel working that reduces the total procurement and delivery time. A further advantage is that documentation is based upon bills of quantities and therefore should be familiar to all concerned. Early price certainty is ruled out, as the client can be vulnerable to any changes in level in the contractor's pricing, between the first and second stages.

Stage one

The first-stage tender is usually based on approximate bills of quantities; however, this does not have to be the case. Other forms of first-stage evaluation may be used; for example, a schedule of rates, although there is perhaps a greater degree of risk associated with this approach. As drawn information, both architectural and structural, is very limited, the choice of bid documentation will be influenced by the perceived complexity and predictability of the proposed project. The two-stage approach places pressure on designers to take decisions concerning major elements of the project at an earlier stage than normally. Assuming that bills of approximate quantities are being used, they should contain quantities that reflect:

- items measured from outline drawings;
- items that reflect the trades it is perceived will form part of the developed design; and
- items that could be utilised during the pricing at second stage.

At an agreed point during the preparation of the first-stage documentation, the design for first stage is frozen, thereby enabling the approximate bills of quantities to be prepared. Without this cut-off point, the stage-one documentation could not be prepared. Note that it is important to keep a register of which drawing revisions have been used to prepare the stage-one documentation for later reference during the preparation of the firm stage two bills of quantities. However, while first-stage documentation is prepared, the second-stage design development can continue. During the first-stage tender period, attention should be focused on the substructure, as it is advantageous if, at the first tender stage, this element is firm. If a contractor is selected as a result of the first-stage tender then they may well be able to start on site to work on the substructure while the remainder of the project is detailed and the second-stage bills of quantities are prepared and priced.

When this procurement route is used, the estimator will have the task of preparing a competitive bid based on limited information. It should also been remembered that the rates included in the first-stage approximate bill of quantities will form the basis for negotiating the second-stage detailed bill of quantities, should the first stage prove to be successful. The estimator should therefore be alert to the temptation of including low rates just to win the first-stage tender.

Once completed, the first stage approximate bills of quantities together with other documentation are despatched to selected contractors with instructions for completion and return in accordance with normal competitive tender practice. On their return, one contractor is selected to proceed to the next phase. It should be noted that selection at this stage does not automatically guarantee the successful first-stage bidder award of the project; this is dependent on the stage-two bidding process.

At this point, the trust stakes are raised. Assuming that a contractor is selected as a result of the stage-one tender, the following scenarios could apply:

Client	Contractor
The client trusts the contractor to be fair and honest during the stage-two negotiations. Failure could result in the client having to go back to the start of the process.	Although selected by stage-one process – there is no guarantee of work. Also no knowledge as to accuracy of first-stage documentation or the client's commitment to continue.
Client could ask contractor to join design team to assure buildability.	Contractor could be asked to start on site on substructure while stage two is progressed.
Client relies on the design team to prepare documentation for stage two timeously. If stage-two bills of quantities are not accurate, the work will have to be remeasured for a third time!	Contractor rewarded on the basis of letters of intent, *quantum meruit*, etc.

The client could, under a separate contract, engage a contractor to carry out site-clearance works while stage-one bids are evaluated.

Contractor could exploit position during stage-two pricing.

Stage two

The purpose of stage two is to convert the outline information produced during stage one into the basis of a firm contract between the client and the contractor, as soon as possible. With a contractor selected as a result of the first-stage process, pressure is placed on the design team to progress and finalise the design. Between contractor selection and stage two, usually a matter of weeks, the design team should prepare and price the second-stage bills of quantities. During this phase, it is usually the quantity surveyor who is in the driving seat and he/she should issue information and production schedules to the rest of the design team. As design work on elements is completed, it is passed to the quantity surveyor to prepare firm bills of quantities, which are used to negotiate the second-stage price with the contractor on a trade-by-trade or elemental basis. It is therefore quite possible that the contractor will be established on site before the stage-two price is fully agreed. Unless the parameters of the project have altered greatly, there should be no significant difference between the stage-one and stage-two prices. Once a price is agreed, a contract can be signed and the project reverts to the normal single-stage lump-sum contract based on firm bills of quantities; however, the adoption of parallel working during the procurement phase ensures that work can start on site much earlier than in the traditional approach. Also, the early inclusion of the main contractor in the design team ensures baked-in buildability and rapid progress on site.

LETTERS OF INTENT

Ideally, before work commences on site a contract should be signed, however there are a number of reasons why this does not happen. Under these circumstances the parties proceed under a letter of intent while they continue to negotiate a full contract. Some letters of intent create binding obligations on the parties and some do not, depending on the circumstances and the drafting. They are not without their problems.

DESIGN AND BUILD AND ITS VARIANTS

Design and Build (D&B), or Design and Construct, is a generic term for a number procurement strategies where the contractor both designs and carries out the works. This approach is extensively used in France, where both contractors and private practices are geared up to provide this service to clients. In the UK, this approach has only become common during the last 30 years or so. The various forms of D&B are described here.

D&B

The contractor is responsible for the complete design and construction of the project. D&B is one of the procurement systems currently favoured by many construction clients. During the past decade, the use of D&B variants in all sectors has increased from 11% in 1990 to over 41% in 2018 (NBS National Construction Contracts and Law 2018; see Figure 1.1). The reasons for this are:

- D&B gives a client the opportunity to integrate, from the outset, the design and the construction of the project;
- the client enters into a single contract with one company, usually a contractor who has the opportunity to design and plan the project in such a way as to ensure that buildability is baked into the design; and
- with specialist involvement from the start, this approach promises a shorter overall delivery time and better cost certainty than traditional approaches.

The results of the studies indicate that D&B outperforms traditional forms of procurement in several respects; however, the differences are not that significant. One reason for this could be that, within the UK, for the organisations that provide D&B services, this is not their key competence and therefore when the opportunity comes to bid for a D&B project, temporary organisations of designers and constructors have to be formed specifically for a project. For the contractor and the designers, the next project may be traditional contracting and therefore the temporary organisation is disbanded. Studies conclude that although delivery times are shorter when using D&B, improvements in cost certainty are only marginal; the total delivery speed of D&B compared with traditional approaches is 30–33% faster. The percentage of projects that exceeded the original estimate by more than 5% was 21% in D&B compared to 32% for traditional procurement.

The main criticisms of D&B procurement are centred around the lack of control over quality of design, with little time being allocated for design development and possible compromises over quality to provide cost savings by the contractor. It is possible for the client to employ independent professional advice to oversee a D&B contract.

Successful use of D&B relies on the contractor preparing proposals that include:

- a contract-sum analysis that itemises the financial detail on an elemental basis; and
- detailed proposals of how the requirements of the client's brief will be satisfied.

When D&B is chosen as the procurement route, the contractor will be responsible for designing, estimating and building the project. It is unlikely that a bill of quantities will be prepared; instead, the contractor will prepare a number of work packages to be priced by sub-contractors in the order required by the project. The process gives more latitude to the contractor to manage the process in a way that maximises profit and delivers the project in the shortest possible time.

For these reasons, tender appraisal can be more difficult when using D&B, as a decision on which tender to accept doesn't only depend on the prices submitted but also the quality of the design and the delivery time. D&B procurement is organised in exactly the same way as single-stage lump-sum procurement, with the drawing up of a short list and bids being submitted to a strict timetable. Other variants of D&B are:

- Novated D&B – the contractor is responsible for the design development, working details and supervising the sub-contractors, with assignment/novation of the design consultants from the client. This means that the contractor uses the client's design as the basis for its bid.
- Package Deal and Turnkey – the contractor provides standard buildings or system buildings that are, in some cases, adapted to suit the client's space and functional requirements.

CONTRACTOR-DESIGNED PORTION

Contractor design is an agreement for a contractor to design specific parts of the works. The contractor may, in turn, sub-contract parts of works to specialists. Some standard forms of contract include an edition to accommodate contractor design, e.g. JCT (16) Intermediate Form of Contract.

MANAGEMENT PROCUREMENT

During the 1970s and particularly the 1980s, commercial clients and property developers started to demand that projects were procured more quickly than had been the case with single-stage selective tendering. The three main management systems are:

- management contracting;
- construction management; and
- design and manage.

With fast-track methods, the bidding and construction phases are able to commence before the design is completed and there is a degree of parallel working as the project progresses. This obviously is high risk as the whole picture is often unknown at the time the works commence on site. This risk is exacerbated when this strategy is used for particularly complex projects or refurbishment contracts. According to the NBS National Construction Contracts and Law Report 2018, management contracting accounts for approximately 1% and construction management 3% of the procurement methods.

MANAGEMENT CONTRACTING

Management contracting is not only popular with developers, as projects are delivered more quickly, but also contractors, as the amount exposure to risk for them is substantially lower than other forms of procurement. This is because a management contractor only commits the management expertise to the project, leaving the actual construction works to others. Management contracting was first widely used in the 1970s and was one of the first so called fast-track methods of procurement that attempted to shorten the time taken for the procurement process. When this procurement method is adopted, the client's quantity surveyor will prepare a number of work-package bills of quantities to be priced by sub-contractors.

Procurement is as follows:

- Selection of a management contractor; as the management contractor's role is purely to manage, it is not appropriate to appoint a contractor using a bill of quantities. Selection, therefore, is based on the service level to be provided, the submission of a method statement and the management fee, expressed as a percentage, of the contract sum. This can be done on a competitive basis. As the management contractor's fee is based on the final contract sum, there is little incentive to exert prudence.
- The management contractor only bids to supply management expertise, although sometimes it can also supply the labour to carry out the so-called *builders' work in connection with services*, as this work package can be difficult to organize.
- The project is divided into work packages. Typically, there are between 20 and 30; for example, groundworks, concrete work, windows and external doors, suspended ceilings, etc. The work packages are, in effect, a series of mini bills of quantities, produced in accordance with NRM2, and therefore the allowed time for pricing an individual package can be reduced to around two weeks. The procedure for asking for bids is the same for single-stage selective tendering. Appendix G of NRM2 gives examples of work-package breakdown structure.
- At the same time, the cost of the project must be determined and therefore an estimated prime cost is established for each package. For the quantity surveyor, this can be problematic as the design is incomplete and therefore estimates of costs tend to be detailed cost plans, agreed with the management contractor.
- Package by package, the works are sent to tender; when a contractor is appointed based on the mini bills of quantities, the contractor enters into a contract directly with the client. Therefore, at the end of the process there are 20 or 30 separate contracts between the management contractor and the work-package contractors; there is no contract between the client and the work-package contractors.
- The management-contractor role therefore is to co-ordinate the work packages on site and to integrate the expertise of the client's consultants.
- Payment is made to the management contractor on a monthly basis, which in turn pays the package contractors in accordance with the valuation.
- The advantages are: work can start on site before the design work is complete; earlier delivery of project and return on client's investment; and the client has a direct link with package contractors.

- In order to provide a degree of protection for the client, a series of collateral warranties can be put in place.
- The disadvantages are: high risk for client; firm price in not known until final package is let; difficult for the quantity surveyor to control costs; and any delay in the production of information by the design team can have disastrous consequences on the overall project completion.
- A JCT (16) Form of Contract exists for management contracts.

CONSTRUCTION OR CONTRACT MANAGEMENT

Construction or contract management is similar in its approach to management contracting in as far as the project is divided into packages; however, the construction manager adopts a consultant's role with direct responsibility to the client for the overall management of the construction project, including liaising with other consultants. Construction managers are appointed at an early stage in the process and, as with management contracting, reimbursement is by way of a pre-agreed fee. Each work-package contractor has a direct contract with the client, this being the main distinction between the two strategies.

DESIGN AND MANAGE

When this strategy is adopted, a single organisation is appointed to both design the project as well as managing the project using work packages. It is an attempt to combine the best of the D&B and management systems. The characteristic are:

- a single organisation both designs and manages;
- the design and management organisation can be either a contractor or a consultant;
- work is let in packages with contracts between the contractor or client, dependent on the model adopted; and
- reimbursement is by way of an agreed fee.

COST-REIMBURSEMENT CONTRACTS

Cost plus

Cost-plus contracts are best used for uncomplicated, repetitive projects such as roads contracts. The system works as follows:

- The contractor is reimbursed on the basis of the prime cost of carrying out the works, plus an agreed cost to cover overheads and profit. This can be done by the contractor submitting detailed accounts for labour, materials and plant that are checked by the quantity surveyor. Once agreed the contractors costs are added.
- There is no tender sum or estimate.

- The greater the cost of the project, the greater the contractor's profit.
- The estimator has little to do in this method of procurement apart from calculate the percentage addition.

Guaranteed maximum price

This method of procurement transfers all risks to the contractor and allows for no increases in price whatsoever, other than costs that result from employer changes. The contractor agrees a guaranteed maximum price to allow for all risks together including risks such as:

- unforeseen ground conditions;
- unexpected encounter with service mains;
- bad weather;
- industrial unrest;
- shortages of labour plant and materials;
- changes in legislation;
- insolvency of suppliers and sub-contractors; and
- fire, storm and earthquake.

Costs are reimbursed on a time and materials basis to a fixed ceiling and the contractor agrees to complete the work without additional payment once the ceiling is reached. In addition, the contractor agrees to accept as payment the lesser of the actual costs of the work plus a fee or the fixed ceiling amount, for example:

- A contractor agrees to build a patio for actual costs plus a £30,000 flat fee, subject to a £100,000 maximum price.
- The contractor will be paid for all costs so long as the total amount, including the fee, does not exceed £100,000.
- If the contractor's costs amount to £80,000, it will not be paid £110,000 (£80,000 in costs plus £30,000 fee), but the £100,000 maximum price.
- Conversely, if the contractor's costs amount to £50,000, it will not be paid the £100,000 maximum price, but £80,000.

Target cost

A variant of cost-plus contracts, this strategy incentivises the contractor by offering a bonus for completing the contract below the agreed target cost. Conversely, damages may be applied if the target is exceeded.

Term contracts/schedule of rates

This approach is suitable for low-value repetitive works that occur on an irregular basic. Contractors are invited to submit prices for carrying out a range of items based on a schedule of rates. Contractors are required to quote a percentage addition on the schedule rates. This method is used extensively for maintenance and repair works.

Negotiated contracts

This strategy involves negotiating a price with a chosen contractor or contractors, without the competition of the other methods. Generally regarded by some as a strategy of the last resort and an approach that will almost always result in a higher price than competitive tendering, it has the following advantages:

- an earlier start on site than other strategies; and
- the opportunity to get the contractor involved at an early stage.

The contractors selected for this approach should be reputable organisations with a proven track record and the appropriate management expertise.

The NBS National Construction Contracts and Law 2018 Report indicates that negotiation is being increasingly used, adding that this may indicate the increased complexity of projects and the increasing willingness to collaborate.

FRAMEWORKS

In both the public and private sectors, frameworks have become increasingly popular. The characteristic of frameworks is that contractors or sub-contractors pre-qualify and then sit on a framework with other selected contractors for a period of time (four or five years). When a client or public-sector body requires a price for a proposed project, a pre-selected framework member is 'called off' the framework to negotiate a price with the client, saving time and tender costs. Importantly for a contractor/sub-contractor, inclusion on a framework is not a guarantee of work and, in addition, some frameworks charge contractors/sub-contractors for both the pre-qualification process and framework membership. Pre-qualification criteria will vary from client to client and are tailored to meet the needs of the sector/client (healthcare, education, etc.).

Pre-qualification

The pre-qualification of contractors and sub-contractors has become commonplace in the UK construction industry. The process involves organisations demonstrating to a client that they have the necessary experience, expertise and culture to be considered to supply a bid for future, but as yet undefined, work.

Registration criteria

Pre-qualification for contractors or sub-contractors may take a number of forms, but the following list gives some indication of typical required pre-qualification information:

Category registration

If the client has a wide variety of work then the pre-qualifying organisations should indicate for which category of work they wish to be considered. For each category, supporting details should be supplied to demonstrate the appropriate staff workforce to carry out the work.

Financial information

In uncertain or difficult trading conditions, when insolvency rates among contractors and sub-contractors are high, it is vital that any organisation selected to carry out a project is financially sound.

If a company is a limited company, a full set of annual accounts, as prepared for shareholders and Companies House, including an audit report or an accountant's report, could be required to be submitted. Any directors, partners and auditors or accountants should sign and date the accounts where necessary. In the case of a group or an ultimate holding firm that does not have consolidated accounts, then accounts of other group firms that are relevant to the application or continued registration could be submitted. These should include:

- a directors' report;
- an auditor's or accountant's report;
- a profit and loss account;
- a balance sheet;
- any notes to the accounts; and
- any supplementary trading account.

In the UK construction industry, the vast majority of contracting organisations are sole traders, partnerships or unlimited companies, and in this case the business's full annual accounts for the latest accounting period, as prepared for the owners and HMRC, including a signed and dated audit or accountant's report, should be submitted along with:

- a profit and loss account; and
- a balance sheet.

References

References from previous employers are also important and should be specific for the work category being applied for. A minimum of two references should be supplied for work recently carried out, say, during the past two or three years. References from previous satisfied clients are one of the most effective ways to judge the suitability of an organisation, provided that they are not for projects such a long time ago that personnel within the organisation may have changed.

Employers' and public liability insurance

Contractors and sub-contractors applying for pre-qualification should be able to demonstrate that they carry effective employers' and public liability insurances. If selected for pre-qualification then evidence of renewal should be demonstrated each year.

Health and safety requirements

Pre-qualifying organisations should demonstrate their health and safety record and this may be done by completing a questionnaire for example, or alternatively current accreditation with a scheme such as Safety Schemes in Procurement (SSIP) could be accepted.

Equal opportunities information

An organisation's policy on equal opportunities should be demonstrated by sub-contractors or contractors. Again, this could be achieved by means of a questionnaire.

Environmental standards

A pre-qualifier's attitude to environmental matters and legislation can be examined by means of a questionnaire; once completed, this can be keep on record.

Geographical area listing

If future projects are nationwide, the pre-qualifiers should indicate the areas of the country in which they are willing to work and the value of contracts they are prepared to take in these areas.

Management team information

Details of the management team together with their qualifications should be submitted for approval.

One of the largest organisations currently operating a pre-qualification database for the construction industry is Constructionline (www.constructionline.co.uk). After a somewhat difficult start, Constructionline is now a robust database of over 20,000 pre-qualified organisations. The pre-qualification criteria have been established and selected by Constructionline, which may or may not correspond to an individual organisation's criteria, but using an organisation such as this obviously saves time and money.

The process

Stage one is the pre-qualification process. It has the following objectives:

- limiting tendering to contractors with the necessary skills and experience to successfully complete the project;
- avoiding unnecessary cost to industry in the preparation of expensive tenders that have limited chance of success; and
- ensuring a competitive tender process, leading to the best value for money outcome for the client.

To achieve this, a six-stage process can be used:

1. Preparation of pre-qualification documents, including a project brief to detail the requirements of the project.
2. Advertisement and issue of pre-qualification documents to interested parties. The documents will require parties to demonstrate various financial, managerial and technical skills in addition to an appreciation for the project. The documents will contain the evaluation criteria, the evaluation procedures and the proposed timing of the evaluation process. Evaluation criteria will be chosen to allow the evaluation team to determine the most suitable parties to be invited to tender. Criteria may typically include:

   - financial status;
   - legal status (entity);
   - relevant experience;
   - available resources (staff, plant, sub-contractor and supplier relationships);
   - performance history, including safety, quality, claims; and
   - demonstrated understanding of the project and associated significant issues including technical, environmental and community.

   Unnecessary and/or irrelevant information and unnecessary copies should not be sought from parties seeking to pre-qualify and should not be supplied.
3. A briefing will be held, at which interested parties will be briefed on the particulars of the project and where parties may ask questions.

4. After receipt of pre-qualifications, submissions should be comparatively assessed in accordance with the evaluation criteria.
5. The evaluation team may seek clarification of any issues from applicants, verbally or in writing, but may not solicit additional information.
6. A list of pre-qualified tenderers is published. Successful and unsuccessful parties should be invited to an individual debrief. When establishing the number of tenderers to be included within the select list, clients should consider the competing aims of:

    - the cost to industry of the preparation of the tender and the possibility of success for any particular tenderer; and
    - ensuring a competitive tender field.

    On a major or complex infrastructure project, the tender list should be restricted to three tenderers. In some circumstances there may be valid reasons to extend the tender list up to a maximum of five tenderers, but in this event the client could contribute to the cost of tendering.

Tender process

The second stage of the procurement process is the tender, which will be put in place with the objectives of:

- allowing each tenderer to develop a design to a sufficiently advanced stage such that it may tender a firm lump-sum offer;
- allowing resolution of general issues requiring clarification to all tenderers; and
- allowing resolution of specific matters only relevant to a particular tenderer's scheme (which require handling with care and strict attention to security).

Information already provided by the tenderer in the pre-qualification process should not have to be provided again in the tender process.

The tender process should be put in place, in eight stages:

1. Tender documents will be prepared by the client, taking into account the issues raised in earlier sections of this model process. The documents will contain the evaluation criteria, the evaluation procedures and the proposed timing of the evaluation process.
2. After issue, an appropriate period will be allowed for the preparation of tenders. It is to the benefit of the client that this period be sufficient to allow tenderers time for the preparation of quality designs, to allow time for innovation and to allow tenderers to minimise risk allowances by finding appropriate alternative solutions. For a relatively simple mid-range project (say £30 million), a period of not less than eight weeks should be allowed; for a major project (say > £65 million), a period of not less than 13 weeks should be allowed.

3. A site visit and a briefing will be conducted, at which tenderers will be briefed on the particulars of the project and may ask questions.
4. If appropriate, a workshop will be held with proponents to discuss particular issues of general interest. For example, a geotechnical workshop may be held, at which tenderers may ask questions of the client's expert and agree on further geotechnical investigations to be carried out by the client.
5. Individual briefing sessions may be held. The objectives of these briefings will be to maintain the confidentiality of the tenderers' intended proposals, while at the same time ensuring that the proposals remain within the client's parameters for the project.
6. After receipt of tenders, submissions are comparatively assessed in accordance with the published evaluation criteria. Tenders are likely to contain significant differences, particularly in the areas of design, time, cost, risk allocation, durability, operation and the like. Tenders may also differ in terms of certainty of delivery and clarity of content.
7. The evaluation team may seek clarification of any issues from applicants, verbally or in writing, but may not solicit new information.
8. A preferred tenderer and a reserve tenderer will be established. Unsuccessful tenderers will be advised as soon as possible. Debriefing meetings will be held with all tenderers.

RULES FOR PREPARING BILLS OF QUANTITIES/WORK PACKAGES

A large percentage of tender documentation is prepared in accordance with rules of measurement principally: *New Rules of Measurement 2* (NRM2) and the *Civil Engineering Standard Method of Measurement 4* (CESMM4) for civil engineering works. A quick glance at NRM2 and CESMM4 may seem to suggest that in terms of format and appearance they are very similar, however the documents have been developed over the years to meet the needs and working practices of very different industries. The essential differences between the two approaches are as follows;

Until the publication of NRM2, the SMM7 (*Standard Method of Measurement*) gave no guidance on the format or preparation of a bill of quantities/work package; it was merely a set of measurement rules. This has been addressed by Part 2 of NRM2, which contains advice (albeit for guidance purposes only) on such things as:

2.2 Purpose of bill of quantities;

2.3 Benefits of bill of quantities;

2.4 Types of bill of quantities;

2.5 Preparation of bill of quantities;

2.6 Composition of a bill of quantities.

In contrast, CESMM4 is more prescriptive in its approach, describing itself as setting out 'a procedure according to which Bills of Quantities shall be prepared'. In the same way, when measuring in accordance with NRM2, the quantity has licence to construct descriptions using the rules as the basis, perhaps adding in non-standard phrases or items, whereas CESMM4 is very specific to point out that work should be 'itemised and described in accordance with the Work Classification'.

SOURCES OF INFORMATION

Where does an estimator find information on rates and prices? There is a number of sources, some more reliable than others. The main sources are:

- A firm's own cost records; these may take the form of previous successful priced bills of quantities or schedule of rates or may be more detailed information on labour outputs and constants for a range of trades and operations. This information is usually closely guarded as it is commercially sensitive.
- Every year around November, a number of builder's price books are published. Perhaps the most well-known is *Spon's Architects' and Builders' Price Book* compiled by AECOM, now in its 143rd edition. Over the past 40 years or so the Spon's *Builders' Price Book* has spawned a number of other versions including *Civil Engineering, Mechanical and Electrical* and *External Works and Landscaping*, priced at between £138 and £160. Many price books are now linked to web-based resources so that pricing data can be kept up to date.
- Estimating software – discussed in Chapter 7.
- The Building Cost Information Service.
- Builders' merchants.
- Sub-contractors.

OPERATIONAL ESTIMATING

Operational estimating is widely used when pricing civil engineering works when it is necessary to consider the overall duration of an operation/operations and the interaction with other trades. One of the fundamental difference between the civil engineering and building contract documentation is that a typical bill of quantities, prepared in accordance with SMM7/NRM2, will be a detailed schedule of items presented in either elemental or trade format. Civil engineering projects, which are always based on approximate bills of quantities that have to be remeasured, tend to have fewer but larger items of, say, bulk excavation or concrete work, and for this reason, from a measurement and estimating perspective, civil engineering has concentrated on operations rather than individual work (bill of quantities) items. Civil engineering projects can be so large that it is unrealistic to concentrate on individual items and therefore this approach is thought to be more realistic (see Table 1.1).

Table 1.1 Sample bill of quantities based on CESMM4

Ref.	Description	Unit	Quantity	Rate	£
	CLASS E: EARTHWORKS				
E421	General excavation; other than topsoil, rock or artificial hard material; maximum depth: not exceeding 0.25 m	m³	120		
E422	General excavation; other than topsoil, rock or artificial hard material; maximum depth: not exceeding 0.25 – 0.5 m	m³	657		
E532	Excavation ancillaries; disposal of excavated material other than topsoil, rock or artificial hard material	m³	267		
E522	Excavation ancillaries; preparation of excavated surfaces; material other than topsoil, rock or artificial hard material	m²	994		
E623	Filling and compaction; embankments; non-selected excavated material other than topsoil or rock	m²	56		

OPERATIONAL APPROACH

Traditionally, when pricing a bill of quantities or work package, the unit-rate estimating approach is adopted; however, this method is not particularly suited with the resource-based information generated by the pre-contract planning process. Therefore an estimator may wish to price some parcels of work, or operations, as a package, rather than a series of separate items or unit rates. For example, the heavy mechanical plant used in earthworks may be used over a range of different bill of quantities items and therefore can be difficult to price accurately merely by calculating a number of individual rates and adding them together. Operational estimating therefore consists of forecasting the cost of a complete construction operation rather than building up rates for the individual items listed in the bills of quantities/work packages. In order to arrive at the total costs, consideration must first be given to what is involved in each element of the operation in terms of quantities, methods and outputs, and these must then be related to the programme.

This is normally done by preparing a method statement, which is a narrative of how the operation is going to be carried out. A method statement should be prepared after an appraisal of the drawings and other relevant data, specification and bill items as well site conditions/constraints and health and safety issues. An operation is defined as a piece of construction work that can be carried out by a gang of operatives without interruption from another gang.

A procedure diagram is provided that shows the relationship of operations to each other and so which are done in sequence and which are done in parallel. This allows a critical-path analysis to be quickly created by the contractor. There are two sections: the first deals with site operations in terms of schedules of materials, labour and plant requirements; and the second, any work prefabricated adjacent or off site. At the end of the bill, management and plant resources are given so they are included.

Between 1960 and 1984, Edward Skoyles was working at the Building Research Establishment on a number of construction industry practices. One of his most high-profile outputs was his work on trying to convince the building industry that it should adopt operational, or as they were later rebadged, activity bills of quantities. When an operational format is adopted, the project under consideration is divided into operations rather than described in terms of the rules of the NRM2.

The advantages claimed for operational bills of quantities included:

- increased accuracy and better cost control; and
- a clearer picture of the resources required to complete an operation and the total costs involved.

But the operational format for building projects never became popular for a number of reasons including:

- operational bills of quantities are very bulky, even for small projects and more costly to produce than traditional formats; and
- the operational or activity-based format does not match the models and conventions in the construction industry for pricing and storing cost data.

Consequently there is little incentive for the building industry to adopt this model and it was never widely used. Even so, operational estimating is not solely confined to civil engineering although its use depends on a detailed understanding of how a particular section of the work will be carried out, for example, lifting precast concrete plank flooring into place.

CIVIL ENGINEERING WORKS

Civil engineering works, by contrast, have taken a broader-brush approach to estimating, due in part to the nature of many civil engineering projects where large quantities are involved as well as extensive use of mechanical plant. In the case of civil engineering works, different approaches to carrying out the works can have a significant effect on prices. All the resources needed for parts of the construction are considered together, instead of in isolation. Examples of where this approach is successfully used are:

- excavation and disposal;
- concrete work; and
- drainage.

There is no hard and fast rule where operational-based estimating techniques stop and unit-rate estimating begins. In fact, it can be difficult to reconcile works priced on an operational basis with a bill of quantities. However, operational estimating is suitable for D&B tendering when the contractor can use its own approach and no bills of quantities have to be submitted.

UNIT-RATE OR RESOURCE ESTIMATING

A price is calculated for each item in the bill of quantities as if the item is to be carried out in isolation to the rest of the works. This is the traditional approach for pricing the majority of building work. The resources required for each bill/work-package item is calculated based on the amount of labour, materials and plant required. This process will be described in more detail in Chapter 4.

ACTIVITY SCHEDULES

An activity schedule is used when a contract is based upon NEC4 Main Options A and C and defined by clause 11.2(20). For option A, an activity schedule is a list of activities prepared by the contractor, which are expected to be carried out in providing the works. When it has been priced by the contractor, the lump sum for each activity is the price to be paid by the employer for that activity. The total of the prices is the contractor's price for providing the whole of the works, including all risks. NEC4 states that:

- information on the activity schedule is not works information or site information;
- the contractor provides information that shows how each activity on the schedule relates to the operations on each programme;
- the prices are lump-sum prices; and
- if the contractor's planned method of working changes so that the activities on the schedule are no longer accurate, the contractor submits a revision.

Although NEC4 is not explicit on which party draws up the activity schedule, it is generally the contractor, with the contractor being paid an amount for each completed activity.

On the one hand, where a bill of quantities is used, the contractor is not at risk if the quantities were incorrectly measured; on the other hand, when an activity schedule is used, the contractor takes responsibility for quantities and pricing.

Table 1.2 Activity schedule

Item no.	Programme reference	Activity description	Price excluding VAT

Therefore, when an activity schedule is used it is the contractor who calculates the quantities and the risk or errors or omissions are transferred to the contractor. It goes without saying that, as the activity schedule is a list of activities the contractor expects to carry out in providing the works, it should correlate with the works included in the works information. As a minimum, an activity schedule should contain the information shown in Table 1.2.

MECHANICAL AND ELECTRICAL WORK

Until the 1970s, it was commonplace for mechanical and electrical (M&E) installations (for example, hot and cold water, lifts, air conditioning, etc.) to be included in the bills of quantities as a prime cost sum to be carried out by a nominated sub-contractor. Note: nominated sub-contractors are no longer appointed when using JCT (16). In the case of some projects, a large proportion of the total cost could be accounted for by way of prime cost sums. Some clients began to question why quantity surveyors should receive fees for work not itemised and measured, and began to demand that M&E items should be measured just like building works. As a result, many practices, largely the big guns of the profession, began to establish their own in-house departments in order to produced detailed bills of quantities for M&E work. M&E specialisation for many quantity surveyors is perhaps regarded as a risky career move when compared with mainstream practice, although in reality M&E quantity surveyors are often in great demand.

CESMM4 requires that bills of quantities include a separate section for method-related charges, which enables the contractor the opportunity to include lump sums relating to the intended method of working. These costs are considered to be proportional to the quantities and are not included in the prices for the individual bill of quantities items. These charges can be adjusted if the engineer instructs the contractor to carry out the work in a different manner as follows:

$$\frac{\text{Original MRC cost} + \text{Additional cost}}{\text{Quantity}} = \text{Rate}$$

METHOD-RELATED CHARGES

Method-related charges (MRCs) refers to a sum inserted into a bill of quantities by a tenderer to cover items or work that relate to the proposed method of construction, the cost of which are not proportional to the quantities and which have not been allowed for in the rates and prices of other items. Each item for MRC shall fully describe so as to define precisely the extent of the work and to identify the resources used. MRCs are either fixed or time-related, as illustrated in Table 1.3.

Civil engineering bills of quantities are provisional and have, therefore, to be remeasured during the actual construction phase.

In summary, MRCs:

- accommodate the variability in civil engineering works;
- describe the cost of performing the bidding contractor's intended method;
- are used at the discretion of the bidding contractor;
- allow the contractor to be paid fairly when quantities change; and
- apply to work whose cost is not directly proportional to its quantity (CESMM4 7.2).

CESMM4 7.1(a) MRCs can be:

- time-related charges; or
- fixed charges.

Both are priced as sums.

The bidding contractor inserts a description of its chosen working method against groups of bill items. Each bidding contractor inserts descriptions of the method to be adopted to include:

- a clear description of what MRCs relate to; and
- a description of plant to be used.

Table 1.3 Method-related charges

Ref.	Description	Unit	Quantity	Rate	£
	GENERAL ITEMS				
	MRCs The contractor is to insert MRCs relating to the works.				

The bill of quantities describes and anticipates a method of construction; however, the contractor may opt to use a different approach. In the case of a variation:

- For a reduction in quantities, MRCs are usually paid in full (CESMM4 7.6).
- For an increase in quantities, costs are adjusted using the contractor's calculation method.

BID PREPARATION

Calculating the true economic cost of carrying out the works, as described in Chapter 4, for a contractor or sub-contractor is only half the bid-preparation process; the second stage is what is known as tender adjudication or tender settlement (Figure 1.2).

Tender adjudication/settlement/director's adjustment

The true commercial cost to a contractor or sub-contractor is calculated from determining the cost of labour, materials and plant but will exclude two vitally important additional components:

- profit; and
- general overheads.

The level of profit enjoyed by contractors/sub-contractors in the construction industry is a closely guarded secret but is generally considered to be low, possibly with the exception of house building, compared with other sectors. However, when considered against the potential risks associated with the construction process, the profit margin or mark-up appears to be almost suicidal. The level of profit will, of course, vary from job to job and from year to year depending on a number of factors as fully discussed in Chapter 5.

BID MANAGEMENT

An important part of preparing a successful bid is the correct management of the process and, in particular, pre-tender planning. Pre-tender planning seeks to:

Figure 1.2 Bid process

- Highlight any critical or unusual activities. Much of what is built is bespoke, that's to say it has a degree of uniqueness and therefore it is important at an early stage in the estimating process to analyse what is to be built and identify the extent of the uniqueness and any unusual details.
- Examine alternative sequences and phasing requirements. Site constraints or client requirements may require site works to be carried out in a non-standard sequence, which may impact on costs. In addition, it may be a worthwhile exercise for the contractor/sub-contractor to examine alternative approaches to site operations to save time and cost.
- Calculate the optimum duration for temporary works and plant to remain in place. Items such as scaffolding can be expensive and therefore should be left in place for the shortest possible time and taken down and removed when no longer required.
- Check whether the completion time is realistic in relation to the complexity of the project, as non-completion may render the contractor/sub-contractor liable for damages.

In addition to the above, there is a number of external factors that can impact on approaches to pricing; these are best considered by addressing the following questions:

- What are we building on – what are the likely site conditions, topography, water table?
- What is to be built – what is the specification of the new project and are new and untried materials and components going to be used?
- Are there any special design factors – is the proposed project a signature building or a straightforward conventional building?
- What impact will regulation, building regulations and health and safety, for example, have on the project?

A vital part of the bid-management process is a site visit by the estimator and the tendering team, and this should be undertaken as soon as possible. As many site photographs and videos as possible should be taken and a site visit report completed as an aide-memoire when back at the office. If a standard site visit report is to be compiled then it should contain provision to record such items as:

- A description of the general locality, together with other local building works currently on site. If there are other building works locally, a visit could be useful if there is the opportunity to investigate current excavations.
- The time of the year and the anticipated weather conditions should be considered; adverse winter weather can impact on trades such as excavation and drainage.
- Ground conditions should be examined together with any signs of surface-water excavations/test pits indicating ground conditions and water table.

- Topographical details should be noted including the position of any trees, including protection orders on the same, and any site clearance or demolition work.
- On brownfield sites, the necessity to decontaminate or remove dangerous materials should be explored.
- The position of the site in relation to road, rail and other transport facilities, as these could impact on getting materials onto the site.
- The availability of space for temporary accommodation should be noted.
- If excavation is to be carried out, the point of the nearest tip for surplus excavated materials should be sourced together with tipping charges; these will have to be taken into account when pricing excavation items.
- The position of existing services and any overhead cables should be recorded.
- If a tower crane is to be used, any restraints imposed by surrounding building, overhangs, etc. should be noted.
- Site access points should be noted together with any restrictions.
- Depending on the nature and location of the project, the need for security should be considered including watching and lighting.
- Finally, local contacts should be recorded, for example, sub-contractors and suppliers.

TECHNICAL REPORTS

During the estimating phase, a number of technical reports and investigations may be produced/available to the estimator including the following.

Site-investigation reports

With an increasing proportion of developments occurring on previously developed brownfield sites and ever more challenging greenfield sites, there has never been a greater need to carry out adequate site investigations. A well-designed site investigation can often lead to project cost savings in the long term by allowing contractors to foresee potential problems. Site-investigation reports range from the simple to the more comprehensive comprising hundreds of pages.

Trial pits

Trial pits can be carried out by a variety of methods from hand-dug pits to machine-excavated trenches. Trial pitting is generally carried out to a maximum depth of 4.5 m with standard excavation plant and, depending on soil conditions, is generally suitable for most low-rise developments.

Window sampling

Window sampling is carried out by either tracked percussive samplers or handheld pneumatic samplers. Samples are retrieved in seamless plastic tubes for logging by a suitably qualified engineer. Window sampling is particularly suited to restricted access sites, contamination investigations and where disturbance must be kept to a minimum.

Rotary boreholes

Rotary drilling techniques are employed where boreholes are required into very dense gravel or bedrock. Samples of bedrock are recovered in seamless plastic tubes for subsequent logging by a suitably qualified engineer and for laboratory testing.

Information about ground conditions will be of importance to a contractor or sub-contractor when tendering on a D&B procurement route as details of ground conditions will impact on foundation design. When a traditional single-stage procurement strategy is used, a contractor will refer to site-investigation reports to determine the water table, for example.

CONSTRUCTION DESIGN AND MANAGEMENT (CDM)

A pre-tender health and safety file is required by the CDM Regulations (2015) and is high on the agenda of tender review meetings, as contractors and sub-contractors need to know the extent of their responsibilities. The estimator may be assisted by the health and safety manager to develop solutions to address such issues as:

- safe access;
- appropriate site facilities;
- environmental issues; and
- Considerate Constructors Scheme.

If, in the opinion of the contractor, the design appears to be unsafe, then the tender submission should be declined. Any cost implications of identified health and safety issues, together with proposed solutions, should be included in the preliminaries and passed onto the sub-contractors for inclusions in quotes. Valuable advice may be obtained from specialist sub-contractors, for example, regarding asbestos removal.

WHAT IS TO BE BUILT?

The materials and components to be used in a project will be described in the specification that accompanies the other tender documents. Generally a specification will take the form of:

- a prescriptive document that spells out in precise terms what materials, components and workmanship standards are required; or
- a performance specification that defines the levels of performance required but leaves it to the contractor or sub-contractors to use appropriate materials or components to meet the performance levels. Even though not prescriptive, the contractor/sub-contractor must still use materials and meet standards defined in codes of practice and British Standards.

The specification will be prepared by the design team using their own procedures, which can vary in coverage and technical coverage. It should be noted that contractors may ignore specifications if they contain standard clauses that are not relevant to the job being priced or are too long. Systems such as NBS (National Building Specification) are now widely used and therefore ensure that specification materials are up to date.

The specification is a contract document and if the tender process is based on drawings and specification, that's to say without a bill of quantities or work package, then the contractor or sub-contractor needs to measure and quantify what has to be priced. This can be done in a number of ways.

INCOMPLETE DESIGN INFORMATION

In a perfect world, all information would be available at tender stage; however, increased pressure on time and professional fees makes this increasingly unlikely. It is therefore probable the information available on which to base the bid is incomplete, and contractors and sub-contractors need to assess the impact of incomplete information on bid strategy. One approach is to include caveats or qualifications within the tender, as these make it clear that certain elements of the proposed project have been excluded from bid. The advantages of this approach are:

- on face value, the bid may appear to result in a lower price than competitors; and
- it mitigates risk due to lack of information.

However, the disadvantages are that

- a bid may be dismissed as incomplete at tender-report stage by the client's quantity surveyor; and
- a contractor/sub-contractor may be asked to remove caveats and supply firm price.

A second approach is to evaluate the impact of the missing/incomplete information, assess the risk posed by incomplete information and put a price on it. The advantages of the approach are:

- the bid will be free from qualifications and easier to assess by the quantity surveyor; and
- it gives the perception of contractor competency.

The disadvantage of this approach is that all risks associated with incomplete information are transferred to the contractor.

OTHER TECHNICAL REPORTS

These may be prepared for:

- special site conditions such as fire risks and security risks;
- the condition of existing structures or temporary works; and
- the treatment of hazardous materials such as asbestos, contaminated ground/ buildings.

METHOD STATEMENTS

Method statements are widely used in the construction industry and they are used in a variety of situations, for example:

- to establish safe systems of work, demolition, alterations and adaptions;
- to investigate an alternative design or approach; and
- to examine the method and resources associated with large-scale activities.

Method statements are descriptions/explanations of how it is intended to carry out certain sections of the works. For D&B tenders, it may be a requirement that a method statement accompanies the contractor's/sub-contractor's bid, and certain clients may require a method statement to be prepared for projects over a certain value as part of their quality-assurance policy. Method statements can be useful for complex refurbishment projects and management contractors may prepare them to plan and manage the levels of interface between work packages. Method statements deal with the use of labour and plant in terms of types, gang sizes and expected outputs.

COVER PRICING

How cover pricing works

In a scenario where three contractors are tendering for a job, one or more of the contractors is chosen by the others to supply a 'cover price' to the other bidding contractors. This cover price is generally artificially high and is submitted to give the appearance of competition rather than with the aim of winning the contract. This allows the other tendering contractors to submit lower bids in order to win the tender.

Whether or not all bidders are involved in collusion, and whether or not the winning bid comes from a colluding party, competition is distorted in all cases. Cover pricing leaves clients unaware that not all the bids they had received are genuine. As a result, they are unable to make an informed decision as to whether to seek replacement bidders, who may have been cheaper.

Because of this, opportunities for other competitors to win the work may also have been reduced. Finally, clients are given a misleading impression of the level of competition available, leading at least potentially to other tender processes being impaired.

Many try to excuse this practice in the situation where a contractor has been asked to submit a bid by a regular client, but is too busy to take on more work. Assuming that refusal to submit a bid will preclude them from being asked in the future, a high cover price bid is submitted, which may or may not be shared with the other tendering contractors.

In April 2008, the Office of Fair Trading (OFT) formally accused 112 construction firms in England of participating in bid rigging on public-sector contracts valued at some £200 million. Some of the largest contractors in the UK, including Balfour Beatty, Carillion and Interserve, were alleged to have participated in cartel-type activity in bidding for public-sector construction contracts, including schools, universities and hospitals. In addition, the OFT alleged that a minority of construction companies entered into agreements where the winner of a contract would make a payment of between £2,500 and £60,000 to the unsuccessful bidder, known as a 'compensation payment'.

In 2010, the magazine *Building* published the results of a survey conducted by Europe Economics for the OFT where 13% of respondents thought that cover pricing was 'common' or 'appears in most bids', which was the same proportion as in 2008, although the National Federation of Builders dismissed the findings. The same survey found a more tolerant attitude by clients to contractors who refused to submit bids by not placing them on a blacklist, an approach recommended by OFT guidelines issued in September 2009.

For the more traditional procurement routes, for example single-stage selective tendering, the estimator will be provided with a bill of quantities or work package that will fully detail the quantities needed. However, increasingly, strategies such as D&B and drawings and specification are being used, and these approaches require the builder or sub-contractor not only to price the job, but also to quantify the nature and extent of materials, labour and plant to carry out the project.

Despite its alleged common nature, bid-rigging is prohibited under the Competition Act 1998 and Article 81 of the European Commission Treaty.

BIDDING STRATEGIES AND MODELS

In its simplest form bidding strategies may be categorized as follows.

Quantity bidding

The most widely used bidding strategy for many contractors is to simply bid on every job that comes along. This high-volume approach is based on the belief that putting out a large quantity of bids means that you will usually win at least a certain percentage of them. This strategy is very time consuming and usually results in low profit margins. The bidding-by-volume approach is most effective for newer companies with little name recognition in the industry that have trouble landing work. It may also be a good strategy for companies struggling to find work, or those that have a large number of employees who are not busy with current projects.

Selective bidding

A more effective strategy is to carefully evaluate bid opportunities based on quality, and to pass on bids that are not a good match for the company. This allows estimators to take their time on each bid and refine their price, which usually results in more successful bids. To utilize this strategy, consider the type of work your company is most successful at. This may be a specific project type, such as hospitals or schools, or a certain size range of jobs. Once you find an appropriate bid opportunity, take the time to produce an accurate estimate and obtain material prices from suppliers. Evaluate the plans and schedule to see how you could perform the job efficiently. This will allow you to keep your bid low and improve your chances of landing the job.

Bidding models

As discussed in Chapter 5, the adjudication process is the final stage in the preparation of a bid and is generally based on:

- the estimator's prediction of the cost;
- an evaluation of the risks and unknowns implicit in the project and the estimating process; and
- the level of profit required to keep the firm solvent and the bank happy.

But how do contractors and sub-contractors know how much the mark-up should be in order to beat the competition? It can be best described as a combination of market intelligence and their perception of how their competitors will behave.

In an attempt to try to introduce more objectivity into the adjudication process, a number of bidding models has been developed over the years with the intention of trying to predict how competitors may approach and decide on the level of mark-up as well as their bidding patterns. These models are based on the premise that

$$bid = estimated\ cost + mark\text{-}up$$

and can be classified as:

- models based on probability;
- regression models; and
- econometric models.

Probability models

As discussed in Chapters 3 and 4, in any given geographical area, say, the M4 corridor or the central belt in Scotland, the costs of labour, materials and plant will be very similar and therefore this approach attempts to analyse the level of mark-up applied by competitors. Using the tender results of a number projects, a competitor's bidding patterns may be determined as shown in Table 1.4.

From these figures, it can be seen that this particular competitor includes mark-up percentages of between 5% and 10% in over 60% of submitted bids. The above exercise can be carried out for other identified major competitors and from the bid frequency distribution a probability curve can be developed illustrating that as the mark-up increases the chance of being the lowest bidder decreases and an optimum bid can be developed.

Regression models

Skitmore and Patchell (1990) have suggested that the use of regression techniques may be used in modelling bidding patterns; however, despite the enthusiasm of the authors, there are so many uncertainties involved with this approach that it is a long way from being a working tool for contractors and sub-contractors.

Econometric models

These models attempt to incorporate the personal feelings and expert knowledge of the bidders in the formal decision-making process. Although bidding models have been around for around 40 years, they appear not to be widely adopted by industry and much of the published information on bidding models dates from the 1980s. This may be because

- markets are volatile and analysis of data using this model does not reflect this;
- contractors and sub-contractors may be happier to 'fly by the seat of their pants' and put trust in hunches; and
- contractors and sub-contractors may not be willing to commit resources to data collection and interpretation.

Table 1.4

Mark-up (%)	Numbers of bids (%)
5	25
10	38
15	20
20	4

2

Early cost advice

In addition to pricing tender documents such as bills of quantities, work packages and specifications, construction clients also require accurate cost advice earlier in the design process as well as an accurate indication on the operating and maintenance costs of the completed project. Generally, the earlier in the design process that figures are required, the more limited the information that will be available on which to base figures; even so, accuracy is still essential. Pre-tender cost advice may be carried out using a number of standard techniques; however, they nearly all depend on calculating the area or footprint of the proposed building/scheme.

RICS PROFESSIONAL STATEMENT: RICS PROPERTY MEASUREMENT 2ND EDITION 2018

For a number of years, UK surveyors have been guided by the Royal Institution of Chartered Surveyors (RICS) *Code of Measurement Practice, 6th Edition* (2007) (COMP) for measuring and calculating the areas of land or buildings. Extracts from COMP are included within NRM1. However, from May 2015, COMP has been gradually replaced by the International Property Measurement Standards (IPMS). Initially, the IPMS were restricted to offices, with residential buildings being added in 2016 and standards for industrial buildings being added in January 2018. IPMS Offices is similar to COMP; however, IPMS Residential is a good deal more complex.

Eventually, IPMS will completely replace COMP. The IPMS coalition of approximately 25 countries is responsible for the initiatives; a full list of countries can be viewed at https://ipmsc.org/.

The move from COMP to IPMS was driven by the lack of common measurement standards and classifications globally. The floor areas of office buildings are measured differently in countries around the world. Jones Lang LaSalle concluded that measurements can vary by as much as 24% and it is hoped that IPMS will:

- introduce consistency of approach;
- capture good practice;
- reduce distortion when comparing cost in different markets; and
- introduce transparency.

The IPMS are split into three sections:

- **IPMS 1**, which equates closely to the former gross external area (GEA) and is the same for all building classes.
- **IPMS 2**, which equates closely to the former gross internal floor area (GIA/ GIFA) and is probably to be the most widely used protocol. IPMS 2 introduces the new concept of internal dominant face (IDF).
- **IPMS 3**, which equates closely to the former net internal area (NIA; also known as net usable area).

One of the major differences between IPMS 2 (Offices) and COMP is that areas that were previously excluded from COMP are now included but identified separately. These include, for example:

- balconies;
- accessible roof-top terraces;
- ground-floor areas not fully enclosed; and
- reveals of windows.

This pocket book will concentrate on IPMS Offices.

IPMS 1 Offices (GEA)

This approach to measurement is recommended for:

- planning purposes; and
- the summary costings of development proposals.

    Calculation:

    Main building: $7.00 \times 12.00 = 84.00 \text{ m}^2$

    Balcony: $3.00 \times 1.50 = 4.50 \text{ m}^2$

Unlike COMP where the GEA would be shown as: $84.00 - 4.50 = 79.50 \text{ m}^2$ when using IPMS 1 Offices, the GEA would be 84.00 m^2 of which 4.50 m^2 would be included in the overall GEA but identified as a separate area of balcony (see Figure 2.1).

IPMS 2 Offices (GEA)

IPMS 2 will be the most widely used IPMS protocol. This approach to measurement is recommended for:

- general-practice surveyors for valuation and lease-agreement purposes; and
- building surveyors and quantity surveyors for preparing cost plans and giving cost advice.

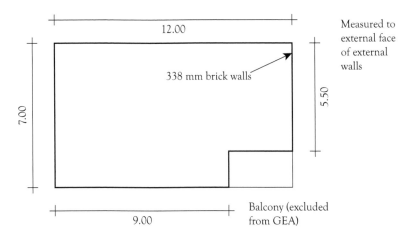

Figure 2.1 IPMS i Offices (gross external area)

IPMS 2 Offices is subdivided into eight colour-coded component areas (i.e. subcategories), A to H:

A Vertical penetrations (e.g. stairs/lift);
B Structural elements (e.g. walls, columns);
C Technical services (e.g. plant/maintenance rooms);
D Hygiene areas (e.g. toilets);
E Circulation areas (e.g. corridors);
F Amenities (e.g. cafeteria);
G Workspaces (e.g. office accommodation);
H Other areas (e.g. balconies).

When using IPMS 2 (Offices) to calculate the floor area shown in Figure 2.2, the area is taken from the internal faces of the external walls (see the following section on IDF):

$5.00 \times 7.00 = 35$ m^2

The difference between COMP and IPMS 2 (Offices) is that the area of the columns will be identified separately as Component area B, Structural elements (e.g. walls and columns) as follows:

Component area B, Structural elements (e.g. walls and columns)   2.16 m^2

$6 \times 0.600 \times 0.600$

Component area G, Workspace (e.g. Office accommodation) $\underline{32.84\ \text{m}^2}$

$\overline{35.00\ \text{m}^2}$

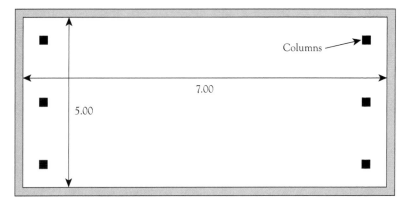

Figure 2.2 IPMS 2 Offices (gross internal area)

IDF/vertical section

In the past, surveyors would measure from above the skirting area to calculate NIA, but to agree an international standard it was necessary to compromise because most of the rest of the world measures from the IDF.

The IDF is the inside finished surface comprising 50% or more of the surface area for *each* vertical section forming an internal perimeter (see Figure 2.3).

Vertical section refers to each part of a window, wall or external construction feature of an office building where the outside finished surface varies from the inside finished surface area of the adjoining window, wall or external construction feature, ignoring the existence of any columns.

The area limit is defined as the position of the IDF for each wall section that makes up the perimeter construction feature. This means that area limits may not

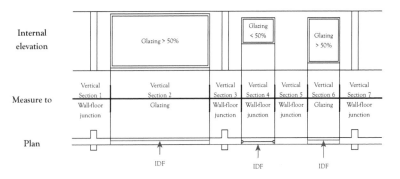

Figure 2.3 Internal dominant face

always coincide with wall–floor junctions but are more than likely to coincide with the face of window glazing. In cases where the internal finishes comprise more than 50% of the floor to ceiling height then the internal finish is deemed to be the IDF. Columns, skirtings, cable ducts, air-conditioning units and cornices are ignored.

Limited-use areas

Limited-use areas have been introduced and are identified and recorded for each component area. Limited-use areas is the term applied to those areas that may warrant special consideration for valuation or other purposes. Guidance is given as to what may constitute limited use:

• IDF and floor–wall junctions;
• areas with limited height;
• areas with limited natural light;
• areas above and below ground; and
• sheltered areas.

In summary

• IPMS 2 Offices is used for measuring the internal area of a building, including internal walls and columns (both were previously excluded from the old GIA).
• IPMS 2 Offices includes (i.e. walls and columns) and excludes (i.e. upper levels of atriums) certain areas from the measurement.
• IPMS 2 Offices includes some measurements that are also required to be stated separately (balconies, covered galleries, roof-top spaces).
• It is recommended that the areas of the various categories are recorded on a pro-forma (included in IPMS 2 Offices) document.

As was the case with quantity surveyors after the introduction of the NRM suite, general-practice surveyors appear not to have welcomed the introduction of IPMS with open arms.

It is mandatory for RICS members or practices to advise clients about the benefits of IPMS, and IPMS should be used unless it is not practical or there are some legislative requirements that preclude its use. IPMS divide buildings into component areas and it is hoped that adoption of the standards and the recommended measurement practice will improve transparency and accuracy.

INTERNATIONAL CONSTRUCTION MEASUREMENT STANDARDS (ICMS)

Along similar lines to IPMS, ICMS are an attempt to establish global consistency in classifying, defining, analysing and presenting construction costs. Published in 2017, ICMS can initially be used for both buildings and civil engineering projects.

It is not the intention to replace existing measurement guidance such as NRM or CESMM; instead, it is a new way of presenting and reporting costs.

THE RICS NEW RULES OF MEASUREMENT 1: ORDER OF COST ESTIMATE AND COST PLANNING FOR CAPITAL BUILDING WORKS

Background

The RICS *New Rules of Measurement* is a three-volume suite of documents that aims to set industry-wide standards for the cost advice, measurement and calculation of whole-life costs.

The first volume, *Order of Cost Estimating and Elemental Cost Planning for Capital Building Works*, was launched in March 2009, with a second edition following in April 2012. Some 390 pages, it aims to provide a comprehensive guide to good cost management of construction projects.

The rationale for the introduction of the NRM1 is that it provides:

- A standard set of measurement rules that are understandable by all those involved in a construction project, including the employer, thereby aiding communication between the project/design team and the employer.
- Direction on how to describe and deal with cost allowances not reflected in measurable building work.

The structure of NRM1 is:

- **Part 1** places rules of measurement in context with the RIBA Plan of Work and the Office of Government Commerce (OGC) Gateway Review* as well as explaining the definitions and abbreviations used in the rules.
- **Part 2** describes the purpose and content of an order of cost estimate and explains how to prepare an order of cost estimate using three prescribed approaches: floor area; functional unit; and elemental method.
- **Part 3** explains the purpose and preparation of elemental cost plans.
- **Part 4** contains tabulated rules of measurement for formal cost plans.

Joined-up cost advice

As a whole, the NRM suite has been developed to ensure that at any point in a building's life there will be a set of consistent rules for measuring and capturing

---

* The OGC was subsumed into the Efficiency and Reform Group within the Cabinet Office several years ago. When it existed, the OGC produced guidance about best practice in public procurement and project management. OGC guidance has now been archived; however, it is still cited in the new Government Construction Strategy. Project planning for public projects continues to follow the OGC Gateway Review.

cost data, thereby completing the cost-management life cycle, supporting the cost management and procurement of construction projects from cradle to grave. The suite comprises:

NRM1 – Cost estimating and cost planning for building works – underpinning how we budget and design buildings.

NRM2 – Detailed measurement for building works – a supporting set of detailed measurement rules enabling work to be bought either through bills of quantities or schedules of rates for capital or maintenance projects.

NRM3 – Rules enabling the measurement of capital cost plans to be integrated with annualised maintenance and life-cycle replacement works.

The NRM has been developed to:

- modernise the existing standards that many of those involved in measuring building work have been used to working with;
- improve the way that measurement for cost planning and bills of quantities has been delivered;
- begin addressing a common standard for life-cycle cost planning and procurement of capital building works, and the life cycle of replacement and maintenance works.

It is important to understand that the NRM is a toolkit for cost management, not just a set of rules for how to quantify building work. As a toolkit, the NRM provides guidance on:

- how measurement changes as the design progresses – from high-level cost/m^2 or cost/functional units to more detailed measurement breakdowns of elements and sub-elements;
- total project costs – it provides guidance on how all cost centres can be considered and collated into the project cost plan;
- risk allowances based on a properly considered assessment of the cost of dealing with risks should they materialise – dispensing with the use of the widely mismanaged concept of contingency;
- total project fees – it provides guidance on how fee and survey budgets can be calculated;
- the suggested design and survey information that a client needs at each RIBA/OGC Gateway stage for the quantity surveyor to be able to provide more certainty around cost advice;
- the suggested key decisions that clients need to make at each RIBA Stage/OGC Gateway stage; and
- a framework for codifying cost plans so they can be converted into works packages for procurement and cost management during construction.

NRM1 is for capital expenditure and NRM3 is for operational expenditure. The Building Cost Information Service (BCIS) has produced a new edition of the Elemental Standard Form of Cost Analysis (SFCA) using the same definitions, coding structure and measurement rules as NRM1 and NRM3. This will allow cost-plan reporting, cost analysis and the benchmarking of capital and maintenance life-cycle cost to use a common format. If clients wish to benchmark costs, then they need to do so on a common basis.

The BCIS SFCA was first produced in 1961 when the bill of quantities was king, was subsequently revised in 1969 and 2008 and has been the industry norm for the last 40 years. In April 2012, to coincide with the publication of NRM1, the SFCA was also updated so that now both the SFCA and NRM1 are in the same format.

Figure 2.4 illustrates a project overview of the NRM from which the various strands of the initiative can be identified.

The RICS formal cost estimation and cost-planning stages in context with the RIBA Plan of Work and OGC Gateway. RIBA Plan of Work is copyright RIBA.

One of the factors that has driven NRM1 is the lack of specific advice on the measurement of building works solely for the purpose of preparing cost estimates and cost plans. As someone who has tried to teach cost planning and estimating for the last 40 years, I am acutely aware that students, as well as practitioners, are often confused as to how estimates and cost plans should be prepared, resulting in the process taking on the air of a black art! This situation has led to an inconsistent approach, varying from practice to practice, leaving clients a little confused. It is also thought that the lack of importance of measurement has been reflected in the curriculum of degree courses, resulting in graduates unable to measure or build-up rates, a comment not unknown during the last 50 years or so.

As illustrated in Tables 2.1 and 2.2, the process of producing a cost estimate and cost planning has been mapped against the RIBA Plan of Work and OGC Gateway process. These show that the preparation and giving of cost advice is a continuous process, that in an ideal world becomes more detailed as the information flow

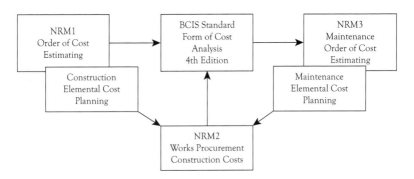

Figure 2.4 Project overview of the NRM

Table 2.1 RICS cost-estimation and planning stages

Design sequence	Process	Advice	Technique	NRM1
0	Strategic Definition	Cost planning	Cost range	Interpolation	Unit method, Elemental or floor area
1	Preparation and Brief	Cost planning	Feasibility study	Interpolation	Unit method, Elemental or floor area
2	Concept Design	Cost planning	Confirm cost limit	Single-rate estimating	Elemental
3	Design Development	Cost planning	Cost plan	Single-rate estimating	Elemental
					Cost planning
4	Technical Design	Cost control	Cost checking	Approximate quantities	Elemental
					Cost planning
5	Construction	Cost control	Cost checking	Approximate quantities	Elemental
				Single-rate estimating	Cost planning

Table 2.2 RIBA Outline Plan of Work, 2013

0	Strategic Definition	Identification of client's needs and objectives, business case and possible constraints on development.
		Preparation of feasibility studies and assessment of options to enable the client to decide whether to proceed. .
1	Preparation and Brief	Development of initial statement of requirements into the project brief or on behalf of the client confirming key requirements and constraints.
		Identification of procurement method, procedures, organisational structure and range of consultants and others to be engaged for the project.
2, 3, 4	Design	**Concept Design**
		Implementation of design brief and preparation of additional data.
		Preparation of concept design including outline proposals for structural and building services systems, outline specification and preliminary cost plan.
		Review of procurement process.
		Developed Design
		Prepare developed design including structural and building services systems, updated outline specifications and cost plan. Completion of project brief.

(continued)

Table 2.2 *(continued)*

		Technical Design
		Preparation of technical design(s) and specifications, sufficient to co-ordinate components and elements of the project. Preparation of detailed information for construction. Preparation of tender documentation
		Identification and evaluation of potential contractors and specialists. Obtaining and appraising tenders.
5	**Construction**	Letting the building contract, appointing the contractor.
		Issuing information to the contractor.
		Arranging handover of site.
		Administration of the building contract to practical completion.
		Provision to the contractor of further information.
		Review of information.
6	**Handover and Close Out**	Handover of building and conclusion of building contract.
		Administration of the building contract and making final inspections.
7	**Use**	Assisting building user during initial occupation period.
		Review of project performance in use.

Source: © RIBA

become more detailed. In practice, it is probably that the various stages will merge and that such a clear-cut process will be difficult to achieve.

NRM1 suggests that the provision of cost advice is an iterative process that follows the information flow from the design team as follows:

- order of cost estimate;
- formal cost plan 1;
- formal cost plan 2;
- formal cost plan 3; and
- pre-tender estimate.

There would therefore appear to be two distinct stages in the preparation of initial and detailed cost advice:

- Estimate = an evolving estimate of known factors. Is the project affordable? The accuracy at this stage is dependent on the quality of the information. Lack of detail should attract a qualification on the resulting figures. At this stage, information is presented to the client as shown in Table 2.3.

- Cost plan = a critical breakdown of the cost limit for the building into cost targets for each element. At this stage, it should be possible to give a detailed breakdown of cost allocation as shown in Table 2.5.

In addition, the NRM approach divides cost estimates and cost plans into five principal cost centres:

1. works cost estimate;
2. project/design team fees estimate;
3. other development/project cost estimate;
4. risk-allowance estimate; and
5. inflation estimate.

The order of cost estimate and cost-plan stages have differing recommended formats (see Table 2.3 for order of cost estimate recommended format). Compared to the

Table 2.3 RICS NRM1 order of cost estimate format

Ref.	Item	£
0	Facilitating Works	
1	Building Works	
2	Main Contractor Preliminaries	
3	Main Contractor Overheads and Profit	
	Works Cost Estimate	
4	Project/Design Team Fees	
5	Other Development/Project Costs	
	Base Estimate	
6	*Risk Allowances:*	
	Design Development Risk	
	Construction Risks	
	Employer Change Risks	
	Employer Other Risks	
	Cost Limit (excluding inflation)	
7	Inflation:	
	Tender Inflation	
	Construction Inflation	
	Cost Limit (including inflation)	

BCIS (SFCA), the NRM format does provide a greater range of cost information to the client, covering the following:

- building works including facilitating works;
- main contractor's preliminaries;
- main contractor's profit and overheads;
- project/design team fees;
- other development/project costs;
- risk;
- inflation;
- capital allowances, land remediation relief and grants; and
- VAT assessment.

RICS NRM – order of cost estimate format

A feature of NRM1 is the detailed lists of information that are required to be produced by all parties to the process; the employer, the architect, the M&E services engineers and the structural engineer all have substantial lists of information to provide. There is an admission that the accuracy of an order of cost estimate is dependent on the quality of the information supplied to the quantity surveyor. The more information provided, the more reliable the outcome will be, and in cases where little or no information is provided, the quantity surveyor will need to qualify the order of cost estimate accordingly.

The development of the estimate/cost plan starts with the order of cost estimate.

Works cost estimate stages 1–3

The works cost estimate has four constituents as shown:

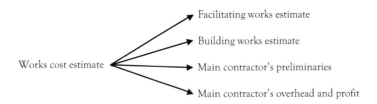

At this stage, the main contractor's preliminaries, overheads and profit are included as a percentage, with sub-contractors' preliminaries, overheads and profit being included in the unit rates applied to building works.

Perhaps the most worthwhile feature of the NRM is the attempt to establish a uniform approach to measurement based on the RICS Code of Measuring Practice

(6th Edition) 2007/IPMS 1, 2 and 3 in which there are three prescribed approaches for preparing building-works estimates, listed here.

Cost per m²of floor area

- GEA – IPMS 1;
- GIA/GIFA – IPMS 2;
- NIA – IPMS 3.

Example

Prepare a works cost estimate for the following proposed scheme given in Figure 2.5.

Figure 2.5 shows a typical floor plan for a new five-storey office block that is proposed to be built in West London. A brief specification has been issued as follows.

SUBSTRUCTURE

Bored piles, reinforced concrete ground beams and pre-cast concrete suspended slab.

FRAME/UPPER FLOORS/STAIRS

In-situ reinforced concrete frame with pre-cast concrete suspended slab/pre-cast concrete stairs.

FAÇADE

280 mm cavity walls rendered externally, with coated double-glazed aluminium windows and external doors.

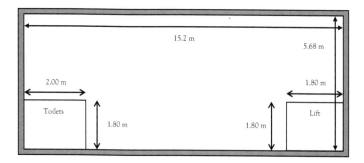

Figure 2.5

INTERNALLY

Hardwood veneered doors, fire resistant to all public and access areas. Plaster and emulsion paint, softwood skirtings, medium quality carpet. Tiling to toilets, bathrooms and WCs. Suspended ceilings throughout.

FITTINGS

White bathroom fittings/mirrors.

HEATING

Electric central heating/air conditioning.

LIFTS

Two-car passenger lift.

EXTERNALLY

Car parking is located adjacent to the development.

Based on the above information, it is possible to produce a building-work estimate based on a cost per m^2 of gross floor area for previous similar buildings.

The GIFA of a typical floor (IPMS 2 Offices: 15.2 m × 5.68 m = 86.34 m^2 split down as follows):

A Vertical penetrations (e.g. stairs/lift)	3.24 m^2
D Hygiene areas (e.g. toilets)	3.60 m^2
G Workspace (e.g. office accommodation)	79.50 m
	86.34 m^2

Five storeys × 86.34 m^2 = 431.70 m^2

From records it is found that the cost/m^2 for an office development of this specification level in Solihull is £1,937.00, although this particular project did not have air conditioning and was completed in 2017. Therefore the price will need to be adjusted for differences in price levels:

	£
Main building	
432 m^2 @ £1,937.00	836,784.00
Add for air conditioning 432 m^2 @ £150 m^2	64,800.00
	901,584.00

Adjust for pricing levels	
Increase in tender levels, say 4%	36,063.36
	937,647.36
Increase for regional variations in price 1%	9,376.47
	947,023.83
Risk allowance 5%	47,351.19
Cost	£994,375.02
Say	£995,000.00

Functional unit method

Building-works estimate = number of functional units × cost per functional unit. A list of suggested functional units is included in the NRM: Appendix B.

Example

An estimate is required for a new secondary school in Hampshire to accommodate 600 pupils. In addition to the standard accommodation comprising class rooms, staff rooms, etc., the local authority wishes to include specialist sports facilities including an all-weather running track. The basic approach when using the unit method is to calculate the number of functional units required for the new project (see Table 2.4), which in the case of a school would be the number of pupils (in this case 600), and multiply this figure by the cost per pupil from a previous similar project.

Although the specifications of the new school and the cost analysis school are similar, the new school is to have more extensive sports facilities and the cost of these must be calculated separately, using a cost/m², as follows:

600 pupils @ £21,000 per pupil =	£12,600,000.00
All-weather running track	£800,000.00
Two football pitches	£1,000,000.00
Two hockey pitches	£800,000.00
Three tennis courts	£1,200.000.00
	£16,400,000.00

Table 2.4 Suggested functional units

Building type	Functional unit of measurement
Car parking	Car-parking space
Hotel	Bedroom
School	Pupil
Hospital	Bed space
Law court	Court room
Prison	Cell
Concert hall	Seat

As straight forward as this sounds, there are certain factors to be taken into account:

- The functional unit cost of a building can vary considerably depending, among other things, on the level of specification and facilities.
- The nature of the functional unit cost makes adjustment for differences between the historical data and the new project problematic.
- The cost of external works, including such items as car parking, running tracks and all-weather pitches, is excluded from the functional unit-rate cost and calculated as separate items and added in.
- It is most suited to preparing early estimates for clients who have a very similar portfolio of property, e.g. supermarkets, retail chains, etc.

As will be discussed in the following example, the functional unit costs must also be adjusted to allow for differences in location and tender levels using an appropriate index.

The elemental method

Of the three methods of approximate estimating recommended by NRM2, the element method is by far the most labour intensive. It is based on the idea that the building-works estimate = the sum of elemental targets and the cost target (for each element) = element unit quantity (EUQ) × element unit rate (EUR). In an ideal world, the following information should be made available to the person preparing the estimate:

- location and availability of the site/site conditions;
- type of building/number and layout of storeys/extent of services;
- statement of floor areas/schedule of accommodation/floor plans/elevations/ sections;
- project/design brief;
- details of any enabling works;
- details of any restraints;
- indicative specification;
- input from structural engineer/M&E engineer, if appointed; and
- initial risk register.

This list, although not exhaustive, is the minimum requirement; however, the more information given to the estimator, the more accurate the final figure.

The amount of detail required to be given with this approach can be seen from Table 2.5, although the choice and the number of elements used to break down the cost of building works will be dependent on the information available. Rules for calculating EUQ are included in Appendix E of NRM1. Bearing in mind that information is still sketchy at this early stage in the design development, if it's impossible to calculate EUQs for a particular element, then the gross floor area method should be used. The recommended major elements commonly used when preparing an order of cost estimate using the elemental method are listed in Table 2.5.

Table 2.5

Group element	Element
0 Facilitating works	
1 Substructure	1.1 Substructure
2 Superstructure	2.1 Frame
	2.2 Upper floors
	2.3 Roof
	2.4 Stairs and ramps
	2.5 External walls
	2.6 Windows and external doors
	2.7 Internal walls and partitions
	2.8 Internal doors
3 Internal finishes	3.1 Wall finishes
	3.2 Floor finishes
	3.3 Ceiling finishes
4 Fittings, furnishings and equipment	4.1 Fittings, furnishings and equipment
5 Services	5.1 Sanitary installations
	5.2 Services equipment
	5.3 Disposal installations
	5.4 Water installations
	5.5 Heat source
	5.6 Space heating/air conditioning
	5.7 Ventilation
	5.8 Electrical installations
	5.9 Fuel installations
	5.10 Lift/conveyor installations
	5.11 Fire/lightning protection
	5.12 Communication, security and control systems
	5.13 Specialist installations
	5.14 Builder's work in connection with services
6 Prefabricated buildings and building units	
7 Works to existing buildings	
8 External works	

Therefore, when sufficient information is available, EUQs are calculated and priced using EURs. Detailed rules for calculating EUQs are contained in NRM1 and currently are still based on COMP. In time, the RICS has stated that COMP will be replaced in NRM1 with IPMS.

In the case where the available information is insufficient to calculate EUQs then the gross floor area method previously discussed should be used.

Even after this process, the data in the cost analysis will need to be adjusted before being used for the cost plan. The front sheet of a cost analysis contains a wealth of information relating to the analysed project that can be used in the adjustment process.

The adjustments to cost-analysis data can be categorised as follows:

- price levels;
- quantity; and
- quality.

DIFFERENCES IN PRICE LEVELS

Differences in price levels between cost analysis and cost-plan data are adjusted using the following:

- building cost indices;
- location indices; and
- tender price indices.

Measures of changes in items such as location, building costs or tender prices are performed using index numbers. Index numbers are a means of expressing data relative to a base year. For example: in the case of a building cost index, a building material is identified, recorded and given an index number. Let's say, for the sake of argument, that the cost of the structural steelwork included in the base index is £90.00 in January 2014. Every three months the costs are recorded for exactly the same materials and any increase or decrease in cost is reflected in the index as follows:

BCIS building cost index; January 2014 = 231

BCIS building cost index; December 2017 = 303

This therefore represents an increase of 31% in the cost of the selected materials, and this information can be used if, for example, data from a 2014 cost analysis was being used as the basis for calculating costs for an estimate in December 2017.

Example

Analysis cost for a steel framed office project (Jan. 2014)   £3,500,000

$$£3,500,000 \times \frac{303}{231} = £4,590,909.00$$

Building cost information such as this can also be used to try to forecast how costs may alter in the future, say during the construction phase of a project.

Any variation in the cost of either of these basics will influence the cost of the works. Cost indices are an attempt to measure price variations that occur between tenders obtained at different times in differing places. The quantity surveyor is able to study the various indices, together with and predicted future cost trends, be it a rise or fall and any regional variations, in order to facilitate adjustments to historical cost information. Indices reflect changes and all indices require the selection of a base period and usually this is set at 100, any increases or decreases being reflected in the indices.

Building cost indices
- The cost of any building is determined, primarily, by the cost of the labour and materials involved in its erection.
- Building cost indices measure changes the cost of materials, labour and plant to the contractor. They ignore any changes in profit levels, overheads, productivity, discounts, etc. and therefore they effectively measure changes in the notional rather than actual costs.
- Building cost indices track movements in the input costs of construction work in various sectors, incorporating national wage agreements and changes in material prices as measured by government index series. They provide an underlying indication of price changes and differential movements in various work sectors, but do not reflect changes in market conditions affecting profit and overhead provisions, site wage rates, bonuses or material price discounts and premiums. In a world of global markets, building cost indices can be influenced by many factors including demands in emerging and developing markets such as China and India.
- Building cost indices are used to adjust and allow for cost increases between the date of the preparation of the estimate and the tender date.

Location indices Tender price levels vary according to the region of the country where the work is carried out. Generally speaking, London and the South East of England are the most expensive and the regional variations are reflected in a location index that is used to adjust prices. The BCIS annually publishes a set of location indices that cover most parts of the UK and these can be used to adjust in cases where the cost-analysis building and the cost-plan building are in different locations.

Tender price indices

- Tender price indices are based on what the client has to pay for a building as they takes into account building costs. These indices therefore reflect fluctuations in the tendering market.
- Tender price indices can be used to adjust for potential increases in cost between the date of the preparation of the cost plan and the actual date the project goes to tender.

OTHER INFORMATION

The front cover of the cost analysis should also contain information relating to the contract type, procurement strategy, market factors, etc. that prevailed at the date the project was current. All of these factors can affect price levels and should be taken into account when preparing a cost plan:

- Contract type: there are a wide variety of contract forms available and within these contract forms, there are a variety of alternatives available. The type of contract used on a project can affect the price and should be allowed for when preparing a cost plan.
- Procurement strategy: similarly, the procurement strategy can affect costs as different strategies will have difference allocations of risk and this will be reflected in price levels, as discussed in Chapter 5.
- Market conditions: when market conditions are buoyant and work is plentiful, contractors may choose to include high profit levels as compared to situations when work is in short supply. Once again, this factor can have an influence of pricing levels and should be taken into account.

DIFFERENCES IN QUANTITY

This adjustment takes account of differences in the elemental quantity of the cost-plan and cost-analysis projects. Table 2.5 shows the information given in an elemental cost plan; EUQs and EURs should be used for this adjustment.

DIFFERENCES IN QUALITY

The final adjustment is an attempt to allow for differences in quality, say finishes, specification levels.

Example – Element 2E – External walls

A cost plan is being prepared for a new six-storey office project with glazed curtain walling. A cost analysis has been selected of a previous similar building and the costs are adjusted as follows:

Data
Cost analysis:
EUR: £1,145.00/m^2
EUQ: 6,100 m^2
Date: Jan. 2015 – BCIS Index: 270
Location: West Midlands – BCIS location factor: 95
Costplan:
EUQ: 5,700 m^2
Date: July 2018 – BCIS Index 314
Location: Yorkshire & Humberside – BCIS location factor: 92

Price levels

Adjust for differences in tender price levels as before:

$$£1,145.00 \times \frac{314}{270} = £1,331.59$$

Adjust for location:

$$£1,331.59 \times \frac{92}{95} = £1,289.54$$

Quantity

The updated cost can now be multiplied by the EUQ for the cost-plan project:

$$£1,289.54 \times 5,700 \text{ m}^2 = £7,350,378.00$$

Quality

The cost-plan building is to have a higher glazing specification than the cost-analysis project:

	£7,350,378.00
Add:	
Self-cleaning glass 5,700 m^2@ £25.00/m^2	£142,500.00
Target cost:	£7,492,878.00

The process is now continued for the remainder of the elements.

Having established the EUQs, the next step is to multiply this figure by the EUR from the benchmark cost analysis. The benchmark rate will almost certainly need to be adjusted for a number of factors including:

- price levels; and
- specification levels.

Example

Benchmark analysis data

The following data are taken from a BICS cost analysis of a previous similar project:

Element 2.2: Upper floors

Pre-cast concrete suspended floor with reinforced concrete topping – £75.95/m^2

New build office building – North Yorkshire

Base date – 15 July 2010

Therefore, applying these data to a similar new project in the South East in December 2018, the following adjustments need to be applied by using a number of published indices.

If the new project is not in the same geographical location as the benchmark project then an adjustment to the rate will be required as the cost of building varies according to location. The adjustment is made by referring to regional price indices: in this case the index for Yorkshire is 99, whereas the index for the South East is slightly higher at 107. The benchmark cost is adjusted as follows:

$$£75.95 \times \frac{107}{99} = £82.09$$

Now this figure should be further adjusted using a tender-based index, such as the one prepared by the BCIS, that will reflect the differences in tender prices between the date of the benchmark project and the preparation of the estimate:

$$£82.09 \times \frac{219}{314} = £117.70$$

This figure can now be multiplied by the EUQ for the new project to give an elemental cost. The rest of the elements should then be calculated in a similar way, although it will almost certainly be the case that some elements will need to be adjusted for difference in specification between the benchmark analysis and the cost estimate.

For example, a benchmark analysis reveals that cost data are based on a project with strip foundations, air conditioning and a lift installation. However, the new project is to have piled foundations and no air conditioning so in this situation cost data relating to these elements should be omitted or revised to reflect the differences in specification. When new cost data are included, they should be compatible with the benchmark material in terms of price levels.

At this stage, main contractor's preliminaries, profit and overheads are recommended to be included as a percentage addition. Sub-contractor's overheads and profit should be included in the unit rates applied to building works.

Project and design fees – 4 In the spirit of transparency, the costs associated with project and design fees are also itemised in the RICS *New Rules of Measurement* formal cost-plan format:

Project and design fees

Project and design team fees

Other specialist consultant fees

Main contractor's pre-construction fees (if applicable)

Main contractor's design fees (where contractor-led deign) (if applicable)

Other development and project costs estimates – 5 This section is for the inclusion of costs that are not directly associated with the cost of the building works, but form part of the total cost of the building project, for example, planning fees.

Risk allowance estimate – 6 Risk is defined as *the amount added to the base cost estimate for items that cannot be precisely predicted to arrive at the cost limit.*

The inclusion of a risk allowance in an estimate is nothing new; what perhaps is new, however, is the transparency with which it is dealt with in the NRM. It is hoped, therefore, that the generic cover-all term 'Contingencies' will be phased out. Clients have traditionally homed in on contingency allowances wanting to know what the sums are for and how they have been calculated. The rate allowance is not a standard percentage and will vary according to the perceived risk of the project. Just how happy quantity surveyors will be to be so up front about how much has been included for unforeseen circumstances or risk will have to been seen. It has always been regarded by many in the profession that carefully concealed pockets of money hidden within an estimate for extras/additional expenditure is a core skill.

So, how should risk be assessed at the early stages in the project? It is possible that a formal risk assessment should take place, and this would be a good thing, using some sort of risk register. Obviously, the impact of risk should be revisited on a regular basis as the detail becomes more apparent.

Risks are required to be included under four headings:

- Design development risks, which may include such items as:
  - inadequate or unclear project brief;
  - unclear design-team responsibilities;

- unrealistic design programme;
- ineffective quality-control procedures;
- inadequate site investigation;
- planning constraints/requirements; and
- soundness of design data.

- Construction risks, for example:

  - inadequate site investigation;
  - archaeological remains;
  - underground obstructions;
  - contaminated ground;
  - adjacent structures (i.e. requiring special precautions);
  - geotechnical problems (e.g. mining and subsidence);
  - ground water;
  - asbestos and other hazardous materials; and
  - invasive plant growth.

- Employer's change risk, for example:

  - specific changes in requirements (i.e. in scope of works or project brief during design, pre-construction and construction stages);
  - changes in quality (i.e. specification of materials and workmanship);
  - changes in time;
  - employer-driven changes/variations introduced during the construction stage;
  - effect on construction duration (i.e. impact on date for completion); and
  - cumulative effect of numerous changes.

- Employer's other risk, a long list of items including, for example:

  Project brief:

  - end-user requirements;
  - inadequate or unclear project brief; and
  - employer's specific requirements (e.g. functional standards, site or establishment rules and regulations, and standing orders).

  Timescales:

  - unrealistic design and construction programmes;
  - unrealistic tender period(s);
  - insufficient time allowed for tender evaluation;
  - contractual claims;
  - effects of phased completion requirements (e.g. sectional completion);
  - acceleration of construction works;
  - effects of early handover requirements (e.g. requesting partial possession);

- postponement of pre-construction services or construction works;
- timescales for decision making.

Financial:

- availability of funds;
- unavailability of grants/grant refusal;
- cash-flow effects on timing; and
- existing liabilities (i.e. liquidated damages or premiums on other contracts due to late provision of accommodation):
  - changing inflation;
  - changing interest rates;
  - changing exchange rates; and
  - incomplete design before construction commences.

Management:

- unclear project organisation and management;
- competence of project/design team;
- unclear definition of project/team responsibilities.

Third party:

- requirements relating to planning (e.g. public enquiries, listed-building consent and conservation-area consent);
- opposition by local councillor(s);
- planning refusal;
- legal agreements;
- works arising out of party-wall agreements.

Other:

- insistence on use of local work people;
- availability of labour, materials and plant;
- statutory requirements;
- market conditions;
- political change;
- legislation; and
- force majeure.

Inflation estimate – 7 Finally, an allowance is included for inflation under two headings:

- tender inflation: an allowance from the period from the estimate base date to the return of the tender; and
- construction inflation: to cover increases from the date of the return of tender to a mid-point in the construction process.

Inflation should be expressed as a percentage using either the retail price index, tender price index or the BCIS building cost indices. This adjustment is, of course, in addition to any price adjustments made earlier in the process when adapting historic cost-analysis data. In addition, care should be taken not to update previous rates that were based on percentage additions e.g. main contractor's preliminaries, main contractor's overheads and profit, and project/design team fees as these will be adjusted automatically when the percentages are applied.

Finally, it is suggested that other advice could be included relating to:

- VAT;
- capital allowances;
- land-reclamation relief; and
- grants.

The extent to which it's possible to give advice on the above will depend upon the size and the expertise of an organisation. In addition, particularly with VAT, the tax position of the parties involved may differ greatly and advice should not be given lightly.

From this point on, advice is given by the preparation of formal cost plans 1, 2 and 3. It is anticipated that for the formal cost-plan stages, the elemental approach should be used and this should be possible as the quantity and quality of information available to the quantity surveyor should be constantly increasing. Table 2.5 demonstrates the degree to which detail increases during this process.

At the formal cost-plan stages, the NRM recommends that cost advice is given on an elemental format and, to this end, Part 4 of the NRM contains comprehensive rules for the measurement for building works and tabulated rules of measurement for elemental cost planning, enabling quantities to be measured to the nearest whole unit, providing that this available information is sufficiently detailed. When this is not possible, then measurement should be based on GIFA. From formal cost plan 2 stage, cost checks are to be carried out against each pre-established cost target based on the cost of significant elements. One thing that's clear is that the NRM approach, if followed, appears too labour intensive and the cost-planning stages and procurement document stages will morph so that the final cost plan becomes the basis of obtaining bids. Over the coming years, it will be interesting to learn to what extent the NRM replaces the tried-and-trusted standard methods of measurement not only in the UK but also in overseas market.

NRM3: ORDER OF COST ESTIMATING AND COST PLANNING FOR BUILDING MAINTENANCE WORKS

Of course, in the modern construction industry not only is an accurate estimate of capital works required by many clients, but also, in addition, estimates are required for maintenance works during the life cycle of the proposed project. To this end,

NRM3 has been devised to provide essential guidance on the quantification and description of maintenance works when preparing initial cost estimates and elemental cost plans. The rules follow the same framework as NRM1. Unlike capital building works projects, maintenance works are carried out from the day a building or asset is put to use until the end of its life. Accordingly, while the costs of a capital building works project are usually incurred by the building owner/developer over a relatively short term, costs in connection with maintenance works are incurred throughout the life of the building – over the short, medium and long term.

For the purpose of developing order of cost estimate, costs in connection with maintenance works, repairs and replacements/renewal works are to be initially ascertained under two separate cost categories as follows:

- Annual maintenance costs, which are divided into the following sub-categories:

  o **Planned preventative costs** – annualised maintenance programme such as preventative maintenance work, includes minor repairs and replacement items (e.g. consumables).
  o **Reactive costs** – annualised unscheduled of responsive maintenance, including minor repairs and replacement items.
  o **Proactive maintenance** – such as planned inspection of buildings, audits, testing/monitoring regimes and specific operation/management procedures.

- Forward renewal works costs, which are divided into the following sub-categories:

  o **Forecast life-cycle renewal plans** – includes cyclical maintenance works (e.g. redecoration and the scheduled major repairs and maintenance works.
  o **Unscheduled repair costs** – e.g. emergency and corrective maintenance.
  o **Unscheduled replacement costs** – emergency/corrective maintenance.
  o **Improvement and upgrades** – as agreed in the project scope.

As with the preparation of an order of cost estimate for building works, the recommended approach by NRM3 is by applying the following approaches:

- cost/m^2 of GIFA;
- functional unit method; and
- elemental format.

NRM3 was published in February 2014 and is effective from January 2015.

3

Resources

The resources on which resource-estimating unit rates are based are, depending on the item being priced, some or all of the following:

- Labour costs – the all-in labour rate. This is built up from operatives' wages plus obligatory statutory 'on costs' such as National Insurance, etc. Labour costs include the costs of employing both skilled (trades) and unskilled (general) operatives together with the additional payments set down in nationally negotiated agreements.
- Material costs – the basic costs of materials plus the costs of delivery, unloading, storage and allowances for wastage.
- Plant costs – the hire cost of mechanical plant such as excavators, dumpers, plus delivery to site, operating costs (drivers and fuel), etc. Some items of plant can be included in the Preliminaries section of the bill of quantities or work package, under the appropriate clause (see NRM2 Clause 2.6.4.1), while other plant costs associated directly with a specific trade or element are included with that item.
- Overheads – overheads include such items as head-office costs, etc. As with plant costs, NRM2 gives the estimator the opportunity to include these costs in the Preliminaries pricing schedule at the front of the bill of quantities or work package. Overheads are in two categories:

  o project overheads, that is costs associated with a specific project; or
  o general overheads, also referred to as Establishment charges; that is, costs associated with meeting the general expenses of the contractor or sub-contractor, to which every contract has to make a contribution. General overheads/establishment charges do not relate to a specific project.

- Profit – the profit margin will vary according to a number of external factors, including market conditions and the perceived risk. In the UK, the profit margin for many general contractors and sub-contractors is surprisingly low and the level of profit recovery is usually determined by the directors during the adjudication process – see Chapter 5.

In a departure from recent practice where contractors may choose to include overheads and profit in the individual unit rates or make suitable allowances elsewhere in the tender, NRM2 requests that only establishment charges (general overheads) are included in the rates with profit and project overheads included in the generally summary as a percentage.

Not all of the above items in a bill of quantities or work package will incorporate all the resources listed. For example, when pricing glazing, there will generally be no need to consider plant costs.

Given that contractors working within a given geographical area will almost certainly obtain their labour, materials and plant from a similar pool, it follows that the competitive/winning edge between competing contractors must be within the costs of overheads and profit. It is therefore important, particularly when market conditions are highly competitive, that overheads are kept to a minimum and contractors and sub-contractors eliminate waste and inefficient practices from their organisations.

UNIT-RATE AND OPERATIONAL ESTIMATING

- Unit-rate estimating: A price is calculated for each item in the bill of quantities as if the item is to be carried out in isolation to the rest of the works. This is the traditional approach for pricing the majority of building work.
- Operational estimating: Civil engineering works, by contrast, have taken a broader-brush approach to estimating, due in part to the nature of many civil engineering projects, where large quantities are involved as well as extensive use of mechanical plant (see Chapter 1).

There is no hard-and-fast rule where operational-based estimating techniques stop and unit-rate estimating begins. In fact, it can be difficult to reconcile works priced on an operational basis with a bill of quantities. However, operational estimating is suitable for D&B tendering when the contractor can use its own approach and no bills of quantities have to be submitted.

Unit-rate estimating

Unit rates will need to be calculated by the estimator for directly employed labour and take into account the following factors:

Labour costs

Labour costs are determined by the calculation of the so called 'all-in' hourly rate and are the basic costs associated with labour with the addition of the costs that comply with a range of statutory requirements. These 'on costs' may include all or some of the following items and can be found in the National Joint Council for the Construction Industry's Working Rule Agreement, which is published annually by the Building and Allied Trades Joint Industrial Council (BATJIC) and available online at: www.fmb.org.uk.

For example, the BATJIC standard rates of wages (plain-time rates) for a 39-hour week were determined as follows for 2018/19:

- City and Guilds London Institute Intermediate/Advanced Craft – between £418.08 and £485.55 per week or £10.72 and £12.45 per hour.
- Adult general operative – £371.28 per week or £9.52 per hour.

Note: these rates change annually and the current rates apply until 23 June 2019.

In addition to the above rates, BATJIC also contains rates for apprentices/ trainees and young adult operatives.

Additional allowances/extra payments for certain duties

Extra payments are made in respect of intermittent or special duties carried out by operatives and are set out within the Working Rule Agreement. The principal on costs are as follows:

WORKING RULE 1C – INTERMEDIATE AND CONSOLIDATED RATES OF PAY FOR SKILL

Additional payments for intermittent responsibility are as follows:

- 47 pence per hour   Air or electric percussion drill, rammer, etc.
- 73 pence per hour   Drag shovel operator, dumper driver, etc.
- 99 pence per hour   Banksman, watchman, pipe-layer, etc.
- £433.74–£456.39 per week   For continuous responsibility for semi-skilled grades, for tower-crane/mobile crane operatives.

In addition, payments are also made on a sliding scale for working at height or in an exposed position.

WORKING RULE 4 – ANNUAL AND PUBLIC HOLIDAYS

Annual holiday amounts to 22 days and public holidays to 8 days.

WORKING RULE 5 – INCENTIVES

Employers and employees may mutually agree incentive schemes for bonus payments.

WORKING RULE 6 – WORKING HOURS

Working hours:

Monday–Thursday 8.00–4.30 pm (4 × 8 hours) = 32

Friday 8.00–3.30 pm = 7

Hours per week = 39

The above is inclusive of a 30-minute lunch break and 10-minute morning break.

WORKING RULE 7 – OVERTIME

The first hour worked over 39 hours is at plain-time rates and thereafter:

Monday to Friday: For the first three hours, time and a half and then double time after that.

Saturday: Time and a half up to 4.00 pm and then double time.

Sunday: Double time.

WORKING RULE 9 – GUARANTEED WEEK

Operatives are guaranteed 39 hours of employment per week despite any stoppages for inclement weather providing that they are available for work.

WORKING RULE 11 – TRAVELLING, FARES AND LODGINGS

There are a variety of daily allowances based on sliding scales to reimburse the costs of travelling to and from work as shown in Table 3.1.

WORKING RULE 12 – SICKNESS AND INJURY BENEFIT

Payment is made to operatives for absence from work due to sickness or injury up to a maximum of 12 weeks per year. The first day of any absence is not paid. Current sickness allowance is £130.00 per week or £26.00 per day and is in addition to statutory sick pay.

Table 3.1 Daily fare allowance

Distance (km)	Fare (£)	Distance (km)	Fare (£)
1–6	Nil	29	7.41
7	0.49	30	7.57
8	0.99	31	7.84
9	1.49	32	7.86
10	1.98	33	8.06
11	2.53	34	8.15
12	3.01	35	8.40
13	3.50	36	8.52
14	4.01	37	8.70
15	4.52	38	8.97
16	4.89	39	9.11
17	5.18	40	9.39
18	5.49	41	9.60
19	5.79	42	9.84
20	5.96	43	10.05
21	6.19	44	10.30
22	6.42	45	10.51
23	6.56	46	10.78
24	6.71	47	10.99
25	6.90	48	11.20
26	7.05	49	11.45
27	7.19	50	11.66
28	7.32		

WORKING RULE 13 – DEATH BENEFIT SCHEME

Payment of £50,000 is made on death of employees over 18 years.

WORKING RULE 18 – TOOL ALLOWANCE

Tool allowance is now consolidated within the basic wage rates.

WORKING RULE 24 – MATERNITY/PATERNITY LEAVE

Employers are bound by statute to provide maternity, paternity leave and adoption leave to eligible employees. Employers are also bound to consider requests for flexible working.

Other allowances/statutory payments

CONSTRUCTION INDUSTRY TRAINING BOARD (CITB) LEVY

This is a levy paid by contractors to the training board to fund the training of new operatives and the development of new skills. The current rate is 0.35% for directly employed labour.

PUBLIC AND EMPLOYER'S LIABILITY INSURANCES

Public liability insurance covers injury to a third party, for example, a passing pedestrian being hit by falling masonry or damage to third-party property. Employer's liability insurance is a legal requirement for limited companies; it covers, for example, the employee who breaks a leg when a trench collapses.

NATIONAL INSURANCE CONTRIBUTIONS

Class 1 National Insurance contributions are payable at the appropriate rate and are calculated as follows. Note that the current threshold for National Insurance payments is £162.00, although this can/does vary:

Crafts

£485.55 weekly pay × 52 weeks =	£25,248.60
Threshold £162.00 × 52 weeks =	£8,424.00
National Insurance 13.80%	£16,824.60 = £2,321.80

General operatives

£371.28 weekly pay × 52 weeks =	£19,306.56
Threshold £162.00 × 52 weeks =	£8,424.00
National Insurance 13.80%	£10,882.56 = £1,501.79

ALL-IN HOURLY RATE CALCULATION

Note all calculations are based on the following:

- City and Guilds London Institute Advanced Crafts earning £485.55 per week or £12.45 per hour;
- adult general operatives earning £371.28 per week or £9.52 per hour; and
- an assumed overtime rate of time and a half – Working Rule 7.

In order to calculate the all-in hourly rate (that is, the hourly rate charged by a contractor for a craft or construction operative), the estimator must first calculate the following.

Hours worked
Total number of hours worked (without holiday allowances):

52 weeks × 39 hours/week	2,028

BATJIC pays 30 days holiday per annum to a total of 234 hours:

17 days annual holiday @ 8 hours/day	136	
5 days annual holiday @ 7 hours/day	35	
7 days public holiday @ 8 hours/day	56	
1 day public holiday @ 7 hours/day	7	
	234	(234)
Standard number of hours worked per year		1,794

Note: It may be that individual organisations reduce this total (based on experience) to cover allowances for bad weather/winter working, etc.

Annual earnings	Craft	General	Craft	General
Basic weekly wages	£485.55	£371.28		
Hourly rate (1/39)	£12.45	£9.52		
Net annual earnings			£22,335.30	£17,078.88
Add				
Guaranteed minimum bonus per hour	£5.00	£2.00	£8,970.00	£3,588.00
Add-on costs				
Non-productive overtime* (time and a half only) say				
300 hours per year			£3,735.00	£2,856.00
Sick-pay allowance @ £26.00 per day (Rule 12)				
Allowance per year (say) 10 days			£260.00	£260.00
Sub-total			£35,300.30	£23,782.88
Add overheads				
National Insurance 13.80% above threshold			£2,321.80	£1,501.79
CITB levy 0.35% of net total pay			£88.37	£67.57
plus holiday pay				

Holidays with pay – 234 hours × hourly rate	£2,913.20	£2,227.68
Death benefit: £9.13 × 12 months plus £7.50	£117.06	£117.06
Sub-total	£40,740.73	£27,696.98
Employer's liability and public liability insurance (say 2% – based on quotation)	£814.82	£553.94
Annual cost of operative to employer	£41,555.55	£28,250.92
Hourly rate – divide by productive hours – 1,794	**£23.16**	**£15.75**

Note that the above rates do not cover costs such as travel or accommodation, administration, supervision etc.

Labour constants

Labour output is the most uncertain part of a unit rate. It can vary considerably depending upon the skills and output of the operative, the site organisation, weather conditions and many other factors often outside the control of the contractor. Historical records of labour outputs are kept by most contractors and sub-contractors based on a variety of sources including observing and benchmarking operations on site. These records, which give the average unit time for each operation, are called labour constants. Rates are usually expressed for an individual craftsman or labourer or, in some cases were more appropriate, for example brickwork, a gang rate may be used. In the case of brickwork this will be built up from calculating the output for two skilled operatives and one labourer who will provide the bricklayers with the materials that they need.

Gang rates

It is often the case that labour rates are calculated on the basis of gang rates rather than individual skilled and labourer's rates. This is thought to be a more realistic approach as often, for example in the case of bricklayers, roofers and plasterers, two or three skilled operatives will be furnished with materials by a labourer, and the gang rate would be based on three skilled operatives to one unskilled operative per hour as follows for bricklaying:

$3 \times £23.16 = £69.48$

$1 \times £15.75 = \underline{£15.75}$

$\underline{£85.23}$ per hour

* Non-productive overtime relates to the additional money paid to operatives working overtime. While working overtime, half as much again in paid and yet the physical amount of work produced is not increased – this unproductive paid working is known as non-productive overtime.

Of course, the output of the bricklaying gang will increase threefold plus compared to an individual bricklayer; economies of scale can also be achieved as a single labourer is used. A variety of approaches have been used in the following examples.

There are a number of other factors that affect the productivity and the output of skilled and unskilled operatives:

- The degree to which incentive and productivity agreements are in place with the workforce. Basically, these schemes incentivise the labour force to increase productivity against pre-determined targets in reward for additional payment. It should be noted that the cost of schemes such as these may not be included in the build-up of the hourly rate, rather they are accounted for separately.
- The hope with incentive agreements is that they will result in greater productivity although in practice the impact of the extra productivity may be cancelled out by the extra cost paid by in incentives.
- A variation on an incentive scheme is a pain-and-gain agreement. This scheme can be introduced between a main contractor and sub-contractor and ensures that if the works are completed more quickly than programmed then any additional profit is shared, whereas if the works finish later than programmed and penalties are incurred then again the financial impact of these (the pain) is also shared.
- In general, production outputs can also be affected by:
  - the skill of the operatives – over the recent past, the tradition of trade apprenticeships has declined; nevertheless, the importance of employing a skilled workforce could result in greater productivity and less work being rejected by the contract administrator and fewer items to be revisited during the rectification period;
  - familiarity with an operation;
  - the quality and appropriateness of the tools and equipment that are available;
  - the quality of the supervision of individual trades;
  - the complexity of the works; and
  - site organisation.
- Programming and planning.

Materials

The bills of quantities require the contractor to price items in terms of units that are, in many cases, unlike the units in which the materials are bought. For example, cement, sand and other constituents of concrete in foundations are bought in a variety of units, such as tonnes, whereas the bill of quantities requires concrete in foundations to be priced in m^3. Therefore, during the estimating process, the builders' merchant's rates and costs have to be converted to the unit required by the bills of quantities.

The all-in rate for materials comprises the following allowances:

- The price quoted by the supplier/builders' merchant. This will have to be adjusted for:
  - ○ trade discounts, these are discounts given by builders' merchants to builders and sub-contractors who have a trade account or can prove that they are a bona-fida contractors; and
  - ○ discounts for bulk orders.
- Transport to the site – mainly included in the quotation price for normal sized items.
- Storage on site – this can vary according to the nature of the materials to be stored and the potential for damage by weather or theft must be considered.
- An allowance for wastage, which will vary according to the type of material but, on average, is about 10%.
- Waste can be categorised into;
  - ○ **Avoidable waste**, such as:
    - – lack of dimensional co-ordination in the design phase leading to materials having to be altered and adapted in-situ;
    - – mistakes by site personnel when ordering;
    - – vandalism and theft;
    - – carelessness and mis-use; and
    - – incorrect storage.
  - ○ **Unavoidable waste**, such as:
    - – breakages in transit; and
    - – cutting to length and size.
- Bill of quantities items are measured net, in other words no allowances are made in the measurement for overlapping on roofing felt of damp-proof courses at joints, see NRM2. An addition has to be made to take account of this.

Plant

As far as a contractor or sub-contractor is concerned, mechanical plant is just part of the stock-in-trade equipment used in day-to-day business and as such can be treated as plant and machinery for accounting and taxation purposes.

For estimating purposes, there are two scenarios relating to the pricing of plant: plant may be owned by the contractor or hired/leased from a specialist plant-hire firm.

Before deciding to own plant, the contractor must consider the cost and the amount of use and work for the machine. Ideally, the plant should be used continuously. It is quite common for larger contracting organisations to have their own separate plant-hire company so that the plant can be used in-house and also hired to other contractors. The principal advantages of hiring plant on an 'as and when' needed basis are that it does not require large sums of capital to be tied up in plant as well as time devoted to maintenance and storage.

From a taxation point of view, both leasing/renting and buying items of mechanical plant have their advantages. As a bona-fida contractor, registered with HMRC, there are a variety of allowances that can be set against liability for taxation either by individuals or companies. These allowances fall into two main categories:

1. capital expenditure, which can be set against taxation at a rate of 18% normally; or
2. revenue expenditure, the more valuable allowance for mainly day-to-day running expenses, which can be set against tax at 100%.

If an item of plant is bought outright then it will be classified for taxation purposes as a capital allowance; however, all lease and rental payments are classified as revenue allowances and therefore are 100% allowable against liability for tax. For example, if a groundworks sub-contractor needs a new excavator at a cost of £20,000, should it be bought or leased?

In the Preliminaries section of NRM2 (1.2.7), there is provision in the Part B Pricing Schedule for the contractor or sub-contractor to include a price for certain items of plant in the bills of quantities or work package. If the section is used, pricing here should be included as a fixed charge and/or time-related charges for the following.

FOR MAIN CONTRACTORS

1. Generally – common-user mechanical plant and equipment used in construction operations for trades such as earthmoving, piling, etc.
2. Tower cranes.
3. Mobile cranes.
4. Hoists.
5. Access plant.
6. Concrete plant.
7. Other plant – small plant and tools.

FOR WORK-PACKAGE CONTRACTORS

1. The type of plant to be provided should be stated with each type separately quantified.
2. Bases.
3. Bringing to site, erecting, testing and commissioning.

4. Dismantling and removing from site.
5. Operator/driver, including overtime.
6. Periodic safety checks/inspections.
7. Other costs/specific charges.

In the case of main contractors, the mechanical plant costs are further sub-divided.

ALTERNATIVE A – BUYING OUTRIGHT

Fixed costs

STRAIGHT-LINE DEPRECIATION

There are two principal approaches for the computation of fixed costs and, in the first scenario, it is assumed that the groundworks sub-contractor is to buy the excavator outright. In this case, the cost of buying the plant will be accounted for with straight-line depreciation; this means that each year for four years, it will be possible to set £8000.00 against a liability for tax. After four years, it will be necessary for the contractor/sub-contractor to purchase a new piece of plant in order to maintain maximum tax relief. If the contractor buys the plant with the use of a bank loan then, in addition to depreciation, the cost of any interest on the loan will also be liable to set against liability for taxation. The following scenario will be repeated for each item of plant bought:

Cost of plant: £40,000

Less scrap value: £8,000 (required by HMRC)

                    £32,000

Assume a life of four years, therefore annual cost is:

$$\frac{£32,000}{4} = £8,000$$

Assume a usage of 1,500 hours per year:

$$\text{Cost per hour } \frac{£8,000}{1,500} = £5.33$$

Writing down depreciation A second approach is to calculate depreciation on a reducing-balance basis, with the item of plant being treated as a capital allowance. This has the advantage of giving maximum tax relief in the early years of the plant's life cycle, utilising the annual investment allowance (AIA).

Purchase price	£40,000	Depreciation
Initial writing down 60% (AIA) during first year	<u>£24,000</u> £16,000	£24,000
Second-year writing down allowance @ 18%	<u>£2,880</u> £13,120	£2,880
Third-year writing down allowance @ 18%	<u>£2,362</u> £10,758	£2,362
Fourth-year writing down allowance @ 18%	£1,936	£1,936
Residual value	<u>£8,822</u>	
Depreciation	£31,178	

As previously: $\dfrac{£31,178}{4} = £7,794.5$

Assuming 1,500 hours per year: $\dfrac{£7,794.5}{1,500} = £5.20$

In both the above cases, the overall hourly rates are similar.

The disadvantage of buying a piece of plant is that it ties up working capital and therefore many contractors prefer to lease machinery over a set period of time. Although the contractor will never own the plant and therefore at the end of the agreement there will be no residual value, the monthly cost of leasing is 100% deductible against tax as it is categorised as a revenue expense.

ALTERNATIVE B – LEASING

If the groundworks sub-contractor decides to lease the new excavator then it will be quoted a fixed monthly payment by the leasing company for a number of years. There is no up-front deposit or capital payment to be made and the amount paid each month is classified as a revenue payment and is therefore 100% allowable against liability for taxation. A word of caution about leasing agreements is that they should not be broken, as the financial consequences can be significant, a point to bear in mind in such a risky business as contracting.

The factors that have to be considered when calculating an all-in rate for plant are:

- Fixed costs:
  - cost of plant and expected operating life;
  - return on capital;

- o maintenance costs; and
- o tax and insurance.

- Operating costs:

    - o operator's wages;
    - o fuel; and
    - o other consumables, including oil etc.

WORKING LIFE

Construction plant comes in a variety of types and usually operates in very extreme conditions; typical values for working life are as follows:

- Concrete mixers 6–7 years
- Cranes 8–10 years
- Dumper 3–4 years
- Excavating plant 5–7 years
- Hoists 5–7 years
- Lorries 3–5 years.

It is usual to assume that plant works for 1,500 hours per year. These figures should be used in the calculation the cost of plant.

OPERATING COSTS

Tax and insurance A road-fund licence is required for plant that uses public highways and generally is insured, for an annual premium, against theft.

Maintenance As with any mechanical item, construction plant will require regular maintenance to work reliably and effectively. Maintenance can be allowed for as a percentage of initial costs, based on manufacturers' recommendations as well as historical records. Allowances vary from 10–30% depending on type.

Operator's wages As mentioned earlier in the chapter, National Working Rule 1C means that operatives with continuous skills or responsibilities are eligible for additional payments. In addition, operators of plant may be eligible for an extra hour per day to fuel and maintain their plant.

Fuel Typical fuel consumptions for items of plant are a matter of record; for example, a rotary drum concrete mixer will use 1 litre of fuel per hour.

Other consumables This item includes oil, lubrication, etc. and can be allowed for by including 20–25% of fuel costs for the item of plant.

Transportation and maintenance The cost of transporting mechanical plant to and from site and the setting up and erection of items such as tower cranes have to be allowed for.

Output The performance of construction plant will affect the cost, and therefore it is essential that know the output or production of a particular item. Once again, these statistics are based on historical records. For example:

- An excavator fitted with a shovel and having a 0.5 m^3 bucket will be able to load 12 m^3 of excavated material into lorries per hour.
- A 14/10 (280 litre) concrete mixer will be able to produce an average of 5 m^3 of mixed concrete per hour.

The output/performance of mechanical plant will be affected by a number of factors, for example:

- site conditions, including time of year;
- the degree to which plant can be incorporated due to restriction on site, site organisation, etc.; and
- skill of the operators

These factors should be taken into account when pricing.

OVERHEADS

Overheads, or establishment charges, are very often difficult to define. For estimating purposes, they include all items that are necessary to the efficient running of a building company and that are not normally charged to a specific contract. Overheads fall into two main categories:

1. Project-specific overheads – these items are usually included within the Preliminaries section of the bill of quantities or work package and can include such items as site accommodation, site security, etc. These are dealt with in Chapter 4.
2. General overheads/establishment charges – these are items such as head-office overheads and salaries, etc. and a percentage must be added to each job in order to contribute to general overheads in order to keep the contractor's organisation in business. The percentage required will vary from organisation to organisation and can be calculated on a previous year's turnover as follows:

Last year's turnover: £30,000,000

Fixed costs: £1,800,000

$$\frac{1,800,000}{30,000,000} \times 100 = 6\%$$

In this example, 6% should be added to the true commercial cost of bill items in order to recover the general overheads of the contractor/sub-contractor.

Profit

The amount of profit required by a building contractor will vary considerably depending on the size of the company, the turnover, market conditions, the contract value and the perceived risks involved. Historically, UK general contractors have operated on profit margins of around 4%, although more for house builders; however, in times of work shortages, contractors frequently tender for work on much less of a profit margin to ensure turnover is maintained. Profit and overheads may be included either as a percentage addition to measured work, in the final summary or in the Preliminaries section, or as a combination of all three.

BREAK-EVEN POINT ANALYSIS

A technique that's sometimes used to establish the trading position and the level of profit being made by a contractor or sub-contractor is a break-even analysis and this is widely used by production management and management accountants. The technique is based on categorising production costs between those that are variable – costs that change when the output changes, and those that are fixed – costs not directly related to the volume of production. Total variable and fixed costs are compared with income in order to determine the level of income at which the business makes neither a profit nor a loss – the 'break-even point'. In its simplest form, the break-even chart is a graphical representation of costs at various levels of activity shown on the same chart as the variation of income with the same variation in activity.

In Figure 3.1, the line OA represents the variation of income at varying levels of activity. OB represents the total fixed costs of the business. As output increases, variable costs are incurred, meaning that total costs (fixed and variable) also increase. At low levels of output, costs are greater than income. At the point of intersection, P, costs are exactly equal to income and neither a profit nor loss is made.

FIXED COSTS

Fixed costs are those business costs that are not directly related to the level of production or output. In other words, even if the business has a zero output or

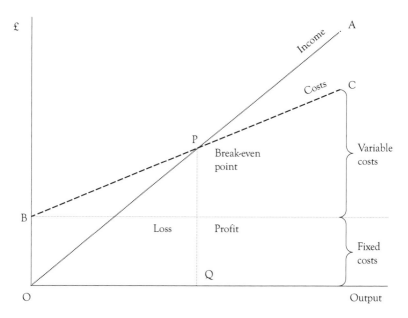

Figure 3.1 Break-even point

a high output, the level of fixed costs will remain broadly the same. In the long term, fixed costs can alter – perhaps as a result of investment in production capacity (e.g. adding a new piece of plant) or through the growth in overheads required to support a larger, more complex business.

Examples of fixed costs are:

- rent and rates;
- depreciation;
- research and development;
- marketing costs (non-revenue related); and
- administration costs.

VARIABLE COSTS

Variable costs are those costs that vary directly with the level of output. They represent payment output-related inputs such as raw materials, direct labour, fuel and revenue-related costs such as commission.

A distinction is often made between "direct" variable costs and "indirect" variable costs.

Direct variable costs are those that can be directly attributable to the production of a particular product or service and allocated to a particular cost centre. Raw materials and the salaries of those working on the production line are good examples.

Indirect variable costs cannot be directly attributable to production but they do vary with output. These include depreciation (where it is calculated related to output e.g. machine hours), maintenance and certain labour costs.

SEMI-VARIABLE COSTS

While the distinction between fixed and variable costs is a convenient way of categorising business costs, in reality there are some costs that are fixed in nature but that increase when output reaches certain levels. These are largely related to the overall 'scale' and/or complexity of the business. For example, when a business has relatively low levels of activity, it may not require costs associated with functions such as human-resource management or a fully-resourced finance department. However, as the scale of the business grows (e.g. output, number of people employed, number and complexity of transactions) then more resources are required. If production rises suddenly then some short-term increase in storage and/or transport may be required. In these circumstances, we say that part of the cost is variable and part fixed.

Example

Turnover of firm		£18,000,000.00
Costs		
Production costs (variable costs)	£14,400,000.00	
Overheads	£2,100,000.00	
Total costs	£16,500,000.00	
		£16,500,000.00
Profit before tax		£1,500,000.00

The break-even point is that at which total income equals total costs. In the example above, it will be the point at which costs added to overheads of £2,100,000.00 will equal the income. If the costs are constant in relation to turnover, the ratio of costs to turnover is:

$$\frac{£14,400,000}{£18,000,000} = 0.8 \text{ or } 80\%$$

The volume of turnover to break-even is:

£2,100.000.00 + 80% of break-even turnover

Therefore £2,100,000.00 = 100 − 80% of break-even = 20%

Therefore breakeven turnover = $\dfrac{£2,100,000.00}{0.20}$ = £10,500,000

Therefore, below a turnover of £10,500,000, a loss will be made. This figure is significant as it will influence the adjudication process discussed in Chapter 5.

4

Unit-rate pricing

This chapter of the book is devoted to examples of building up unit rates from basic principles, referred to throughout this book as resource estimating. The examples are based on the RICS NRM2, although the techniques used equally apply to other codes of measurement. Note that the enhanced labour rates used in the following examples were calculated previously in Chapter 3.

PRELIMINARIES

A contractor/sub-contractor is not obliged to price the Preliminaries section of a bill of quantities or work package, and some organisations may choose to split the cost of providing the preliminaries items between the Preliminaries section and the rates. However, many will choose to price the Preliminaries section. One of the main reasons for adopting this approach is that in the case of an extension of time being granted by the contract administrator, the contractor/sub-contractor may be entitled to addition payment to cover preliminary items for the extended period. Having the preliminary cost spilt into fixed and time-related costs makes the process much easier.

In addition, the schedule of preliminaries pricing is used in the calculation of monthly payments to be included in interim valuations. It is important that contractor/sub-contractor prices the Preliminaries section accurately as detailed below and not simply apply for payment on a proportional basis. As the site set-up costs at the start of a contract can be high and recovery is by way of monthly valuations, any inaccuracies in the estimation of preliminaries items will result in an under-recovery of costs at the start of a contract that may take several months to rectify itself and have a detrimental impact on cash flow.

NRM2 contains a pricing schedule for preliminaries in Appendices C and D, and it is recommended that this schedule should be priced and submitted by the contractor with the completed tender. In addition, NRM2 recommends that a 'full and detailed breakdown' should be submitted at the tender stage, but perhaps this is asking for a little too much.

The first section in a bill of quantities or work package is Preliminaries. It is recommended by Work Section 1 of NRM2 that Preliminaries are described and quantified along the following lines.

The Preliminaries section (main contract) is divided into two sub-sections:

- Part 1: information and requirements (i.e. dealing with the descriptive part of the preliminaries); and
- Part 2: pricing schedule (i.e. provides the basis of a pricing document for preliminaries).

The pricing schedule for work-package preliminaries is divided into two main cost centres:

- employer's requirements; and
- main contractor's general cost items.

Work-package contract preliminaries

As with the main contractor's preliminaries, the work-package preliminaries are split into:

- Part 1: information and requirements (i.e. dealing with the descriptive part of the preliminaries); and
- Part 2: pricing schedule (i.e. provides the basis of a pricing document for preliminaries).

The pricing schedule for work-package preliminaries is divided into two main cost centres:

- employer's requirements; and
- work-package contractor's general cost items.

Fixed and time-related charges (main and work-package contractors)

NRM2 splits Preliminaries general cost items into:

- **fixed charges** – the cost of which is considered to be independent of duration, that's to say charges that are not proportional either to the quantity of the work or its duration, for example:
  - ○ temporary water-supply connection; and
  - ○ licences in connection with hoarding, scaffolding, gantries and the like; and

- **time-related charges** – the cost of which is considered to be dependent on duration, that's to say charges that are directly proportional to either the quantity of the work or its duration, for example:

  o cleaning; and
  o general office furniture, including maintenance.

Although the Preliminaries section of bills of quantities tends to contain many items, several of them are difficult to quantify and price, and for this reason it is usually only the major items such as accommodation, staffing and mechanical plant that are priced. Appendices to NRM2 give templates for preliminaries pricing schedules.

The main difference between the Preliminaries section of main contractors and work contractors is that the amount of detail is somewhat reduced and tailored to the specific work package in the latter, although some items in the work-package preliminaries refer back to the main contract for the level of detail required.

Example

Part 1 – employer's requirements: site accommodation/services/facilities/ temporary work (1:11)

Site accommodation/site offices/information and requirements

The contractor shall provide suitable accommodation for the Clerk of Works, with a minimum floor area of 20 m^2 including a desk, four chairs, a filing cabinet and all necessary heating, lighting and cleaning.

Part 2 – pricing schedule

It is recommended that this item is priced as follows in a pricing schedule:

Cost centre	Component	Time-related charges (£)	Fixed charges (£)	Total charges (£)
1	Employer's requirements			
1.1	Site accommodation			
1.1.1	Site accommodation	20,280.00	636.25	20,916.25
1.1.2	Furniture and equipment	6,400.00	0.00	6,400.00

Data

Contract period: 160 weeks   Transport to and from site: £200 per trip
Hire of hut: £90 per week   Hire of furniture and heaters, etc.: £40 per week
Energy costs: £5 per week

Fixed charges:	£	£
Transport to and from site (2 trips × £200)	400.00	
Erection 10 hours		
Dismantle <u>5 hours</u>		
15 hours labourer @ £15.75	<u>236.25</u>	
Total of fixed charges	<u>636.25</u>	636.25

Time-related charges:		
Hire of hut 160 weeks @ £90 per week	14,440.00	
Furniture and heaters 160 weeks @ £40 per week	6,400.00	
Energy costs for heating & lighting		
160 weeks @ £5.00 per week	800.00	
Cleaning – 2 hours per week × 160 weeks =		
320 hours @ £15.75	<u>5,040.00</u>	
Total of time-related charges	26,680.00	<u>26,680.00</u>
Rate per week = £170.73		
Total (Fixed and time-related charges)		<u>£27,316.25</u>

Example

Part 1 – employer's requirements: security, safety and protection (1.9)

Hoardings, fences and gates

The contractor is to allow for enclosing all boundaries of the site. It shall be the contractor's responsibility to provide all necessary precautions, protection and security to safeguard the works using fences, hoardings, gates, etc. as considered necessary.

Part 2 – pricing schedule

Cost centre	Component	Time-related charges (£)	Fixed charges (£)	Total charges (£)
2.4	Security			
2.4.3	Hoardings, fences and gates	4,670.00	1,200.00	5,620.40

Data
Contract period: 160 weeks

Fencing £35 per m	£	£
Fixed charges; erection and taking down		1,200.00
50 m of solid fencing 1.8 m high @ £35/m	1,750.00	
1 pair gates and security	<u>400.00</u>	
	2,150.00	2,150.00

Maintenance, repairs and adaption:

1 hour per week labourer – 160 weeks @ £15.75	2,520.00
Total	£5,870.00

Example

Some items of mechanical plant are used by several trades (common user) and as such it is difficult to apportion the cost accurately to each trade; therefore the contractor has the opportunity to allow for them in the Preliminaries section. Typical items included for are tower cranes, hoists, etc.

Part 1 – employer's requirements: site accommodation/services/facilities/temporary work (1:11)

Common-user mechanical plant and equipment used in construction operations.

Part 2 – pricing schedule: mechanical plant

Cost centre	Component	Time-related charges (£)	Fixed charges (£)	Total charges (£)
2.7	Mechanical plant			
2.7.1	Generally	2,446.78	1,500.56	3,947.34
2.7.2	Tower cranes	9,240.48	10,300.00	19,540.48
2.7.3	Mobile cranes	7,060.00	Nil	7,060.00
2.7.4	Hoists	500.89	123.45	624.34
2.7.5	Access plant	458.77	345.00	803.77
2.7.6	Concrete plant	3,000.98	450.00	3,450.98
2.7.7	Other plant	200.32	198.56	398.88

The type of plant to be provided shall be stated, with each type separately quantified. It is recommended that the following items should be included:

1. bases;
2. bringing to site, erecting, testing and commissioning;
3. dismantling and removing from site;
4. protection-systems item;
5. operator/driver, including overtime week (number of staff by number of man hours per week by number of weeks);
6. periodic safety checks/inspections monthly; and
7. other costs' specific charges item.

Example

50 TL GFS tower crane.

Cost centre	Component	Time-related charges (£)	Fixed charges (£)	Total charges (£)
2.7	Mechanical plant			
2.7.2	Tower cranes	13,773.54	10,300.00	24,073.54

Data

Contract period: 6 weeks
Plant charges – crane
Base: £300
Crane hire @ £500/week for 6 weeks
Crane transport to site and from site: £5,000 each way

Fixed charges:		£	£
Base to crane		300.00	
Transport to and from site	2 trips × £5,000	10,000.00	
Erection	Included in above		
Dismantle	Included in above		
Total of fixed charges		10,300.00	10,300.00

Time-related charges:

Hire of crane 6 weeks @ £1,000 per week	6,000.00

Labour

Driver – 37 hours per week @ £23.16 = £856.92		
£852.92 plus £442.67 WRA = £1,295.59 × 6 weeks =	7,773.54	
Total of time-related charges	13,773.54	£13,773.54
Total (Fixed and time-related charges)		£24,073.54

BUILDING WORK

SECTION 2: OFF-SITE MANUFACTURED MATERIALS, COMPONENTS OR BUILDINGS

Section 2 of NRM2 gives the opportunity to include items that are complete or substantially complete building elements of proprietary construction, largely prefabricated.

It recognises that increasingly parts or even whole buildings are prefabricated off site and brought to site for assembly. Bathroom pods for inclusion in hotels or halls of residence are an example of this.

The fixing of items supplied only as part of the proprietary package is included and priced here. Other work not forming part of the proprietary package is measured separately in the appropriate work section.

Most items will be enumerated, and when pricing these items it should be remembered that NRM2 states that the following items are deemed to be included; in other words, the estimator must include them in the bill/work package rate:

- all factory-applied finishes;
- transport from factory to site;
- off-loading and storing on site;
- setting, hoisting and placing in final position;
- all connection and joint materials;
- all service connections; and
- disposal of all packaging and protective materials.

If the off-site manufactured units are being provided by a sub-contractor, either domestic or named, the estimator should ensure that the above list of items has been allowed for somewhere in the pricing, if appropriate.

SECTION 3: DEMOLITION

Even for large main contractors, any demolition work will be carried out by a specialist contactor, due to the high risk and insurance premiums associated with this type of operation. The cost will vary enormously depending on the size and location of the work and is usually priced on the basis of a cost per cubic metre. This section also includes items such as decontamination and the recycling of existing materials.

SECTION 4: ALTERATIONS, REPAIRS AND CONSERVATION

NRM2 gives this section a certain amount of importance with a number of items included. Work in this section is generally a combination of trades and the price will be based on rates calculated in other work sections, although by its very nature alteration and repair work cannot be carried out at the same rate as new build and the rates may have to be adjusted accordingly.

Example

Works of alteration

Alter and adapt existing opening in one-brick wall plastered one side where hatch removed (measured elsewhere) to form new opening size 1.20 × 1.20 m including extending one jamb for a width of 0.40 m and a height of 1.20 m in 225 mm brickwork in common bricks in gauged mortar (group 3), cut tooth and bond new to existing, wedge and pin at head, extend finishes and make good all work disturbed. Cost per item.

The first item that the estimator needs to do when pricing an item similar to the one above is to split the work down into its constituent parts, calculate quantities where necessary, price each part separately and the add up the results.

- One-brick wall in common brickwork: $0.40 \times 1.20 = 0.48$ m^2
- Cut tooth and bond new to existing: 1.20 m
- Wedging and pinning at head: 0.40 m
- Extending plaster both sides (assumed that decoration taken elsewhere) = 0.48 m$^2 \times 2 = 0.96$ m^2
- Making good all work disturbed.

Build-up   £

0.48 m^2 one-brick wall in common brickwork, based on £102.62/m^2 49.26

1.20 m cut tooth and bond @ 2 hours bricklayer and
   1 hour labourer/m

        £46.32
        <u>£15.75</u>
   1.20 × £62.07   74.48

0.40m wedging and pinning @ 0.40 hours bricklayer and
   0.20 hours labourer/m

   0.40 × £23.16   9.26
   0.20 × £15.75   <u>3.15</u>
                £12.41 × 0.40   4.96

0.96 m^2 two-coat plaster @ £26.51 (from p. 158)   25.45
Making good all work disturbed, say £5.00   <u>5.00</u>
                                             159.45
Add profit and overheads 15%   <u>23.92</u>
Cost per item   <u>£183.37</u>

Example

Take down 230 mm wide hardwood capping to existing blank opening and alter and adapt to form opening to the new first-floor extension including cutting away one-brick wall, plastered one side for a width of 3 m and a height of 0.75 m (to slab level), including quoining up jambs, extending finishes and making good threshold and all work disturbed. Cost per item.

- Take down capping
- Cutting away one-brick wall: 3 m × 0.75 m = 2.25 m^2
- Removing spoil from site: item
- Quoining up jambs: 0.75 m
- Making good threshold: 3 m
- Make good all work disturbed

Build-up £

Taking down capping – say 5.00

2.25 m^2 cutting away one-brick wall @ 0.40 hours bricklayer and
0.20 hours labourer/m^2

0.40 × £23.16
0.20 × £15.75 = £12.41 × 2.25 m^2 27.92

Removing spoil 8.00

0.75 m quoining up jambs @ 0.90 hours bricklayer and
0.40 hours labourer/m

0.90 × £23.16
0.40 × £15.75 = £27.14 × 0.75 m 20.36

3.00 m making good threshold
1 hour bricklayer @ £23.16 23.16
Make good all work disturbed, say <u>5.00</u>
89.44
Add profit and overheads 15% <u>13.42</u>
Cost per item <u>£102.86</u>

GROUNDWORKS

SECTION 5: EXCAVATION AND FILLING

An important point to remember with groundworks is that when ground is excavated, it increases in bulk. This is due to the fact that broken up ground has a greater volume that compacted ground; a point that has to be taken into consideration when transporting or disposing of excavated materials. Volumes in the bills of quantities are measured net, without allowance for bulking; therefore the estimator must make an allowance. What's more, the degree to which excavated material bulks depends on the type of ground as follows:

- Sand/gravel: 10%
- Clay: 25%

- Chalk: 33%
- Rock: 50%

Other points to take into account are:

- The time of year when the excavation is to be carried out and whether the excavated materials and the bottoms of the excavation are likely to be flooded.
- The nature of the ground will affect the type of earthwork support required and the labour constants for carrying out the works. Earthwork support is often regarded by contractors as a risk item and as such may not actually be used although priced in the bills of quantities. For this reason, in NRM2 earthwork support is not measured unless specifically requested.
- The distance and availability of the nearest tip to which any excavated material has to be transported plus charges for tipping. Contaminated spoil must be dealt with and disposed of separately, and higher tipping charges will be incurred for contaminated waste.
- The sections of the works that are to be carried using hand digging or mechanical plant.

It is advisable to visit the site, study the results of test pits, soil samples, etc. and to check site access.

Hand excavation or machine?

The majority of excavation is now done by machine; however, there may be instances where there is restricted access, or where there are small quantities involved, where hand excavation is used. The unit rate for hand excavation is generally more expensive than excavation by machine.

The following are average labour constants for hand excavation under normal conditions and in medium clay or heavy soil:

Item	Hours/m^3
Bulk excavation commencing at reduce levels not exceeding 2 m deep	2.40
Ditto over 2 m not exceeding 4 m deep	2.70
Foundation excavation commencing at reduce levels not exceeding 2 m deep	4.00
Bulk excavation, commencing at reduce levels not exceeding 2 m deep	3.30
Filling obtained from excavated materials final thickness exceeding 500 mm deep	1.00
Retaining excavated material on site in temporary spoil heaps, distance not exceeding 50 m	1.00

For excavation other than medium clay, the following multipliers should be applied to the above:

Loose sand: 0.75

Stiff clay or rock: 1.50

Soft rock: 3.00

Hand excavation

Example

Bulk excavation commencing at reduce levels not exceeding 2 m deep. Cost per m³.

Assume 10 m³	
24 hours labourer @ £15.75	£378.00
Add profit & overhead 15%	£56.70
Cost per 10 m³	£434.70
÷ 10 – cost per m³	£43.47

Example

Foundation excavation in trenches, commencing at reduce levels not exceeding 2 m deep. Cost per m³.

Assume 10 m³	
33 hours labourer @ £15.75	£519.75
Add profit and overheads 15%	£77.96
Cost per 10 m³	£597.71
÷ 10 – cost per m³	£59.77

As mentioned previously, it is uncommon to carry out excavation by hand unless the work is restricted to the extent that there is no room for machinery to operate. Before pricing any of the items involved, the estimator must decide on the type or types of machines to be used and the respective size of each machine. The following is a guide to the correct type of machine for each operation. The size of the machine will largely depend on the quantities of excavation involved.

Scraper

Used for large-scale over-site excavation to remove vegetable soil and spreading excavated material on site (Figure 4.1).

Figure 4.1 Scraper

Source: iStock

Excavator with backactor

Used for trench and basement excavation where the machine has to stand at ground level (Figure 4.2).

Figure 4.2 Excavator with backactor

Source: Photo © Craig Maybery-Thomas. From online resources for *Fundamental Building Technology* (2012)

Excavator with shovel

Used for bulk excavation and basement excavation where the machine can stand in the basement to work as well as loading excavation into trucks for removal (Figure 4.3).

Backactors and excavators are available with a variety of different sized shovels and the correct sized shovel should be used to ensure optimum output and efficiency. Finally, the cost, if applicable, of cleaning roads should be considered, taking into account the expected weather conditions and time of year during which the machinery will be operating.

Example

Foundation mechanical excavation in trenches, commencing at reduce levels, not exceeding 2 m deep. Cost per m³.

Data

The use of 0.25 m³ bucket excavator @ £50.00 per hour including labour and consumables. The output for a 0.25 m³ bucket excavator is 5 m³ per hour.

Note when using mechanical plant for excavation, a banksman is usually required who is paid the labourer rate plus a small addition for extra responsibility. The banksman is responsible for making sure that the excavation is carried out to the correct depths.

Figure 4.3 Excavator with shovel
Source: iStock

Assume 10 m³

Plant:	
2 hours excavator @ £50.00 per hour	100.00
Banksman:	
2 hours banksman @ £15.75 + 0.96 (WRA)	<u>33.42</u>
	133.42
Add profit and overheads 15%	20.01
	<u>153.43</u>
÷ 10 – cost per m³	<u>£15.34</u>

Disposal

Example

Disposal of excavated materials off site. Cost per m³.

Assume 6 m³
Tipping charge £10.00 per load
Site to tip 4 km round trip
6 m³ lorry @ £40 per hour including driver and fuel
Bulking of excavated material 33%

$$\text{Load as dug } 6\text{m}^3 \times \frac{133}{100} = 4.5\text{m}^3$$

Time per load (minutes):	
Assume average speed of lorry 30 kph	
1 kilometer in 2 minutes	
4 kilometers travel = 4 × 2 = 8 minutes	8.00
Time to load lorry	
Excavator bucket size 0.33 m³	
Cycle time per bucket 0.5 minutes	

$$\frac{4.5}{0.33} \times 0.5 = \qquad\qquad 6.82$$

Tipping time	<u>4.00</u>
Total operation time	<u>18.62</u>

$$18.62 \text{ minutes} \times \frac{40}{60} \qquad\qquad 12.41$$

Tipping charge	<u>10.00</u>
Cost per 4.5 m³	<u>22.41</u>

Cost per m³	4.98
Add profit and overheads 15%	0.75
Cost per m³	£5.73

The output of machine excavators will depend on the size of the machine, the method of disposal, and the accuracy of the finished work, as well as the skill of the operator. Typical times based on a 0.25 m³ bucket are as follows:

Operation	Hours/m³
Excavation to reduce levels	0.10
Excavation for basements	0.15
Excavation for trenches	0.20
Excavation for pits	0.25

Landfill tax

When disposing of excavated materials the cost of landfill tax must also be considered. Landfill tax is a tax on the disposal of waste that is administered by HRMC. It aims to encourage waste producers to produce less waste, recover more value from waste, for example through recycling or composting, and to use more environmentally friendly methods of waste disposal. The tax applies to all waste:

- disposed of by way of landfill;
- at a licensed landfill site;
- on or after 1 October 1996;
- unless the waste is specifically exempt.

The tax is charged by weight and there are two rates. Inert or inactive waste is subject to the lower rate. The standard rate of landfill tax was £88.95 per tonne from 1 April 2018. The lower rate of landfill tax (which applies to 'inert' waste only) was £2.80 per tonne from 1 April 2018. The measure will affect those who dispose of waste to landfill, including local authorities and, ultimately, individuals who pay the associated council-tax charges.

Fillings

Hardcore filling

Materials used as hardcore, such as brick, stone, etc. are bought either by volume or weight. If bought by volume, an average of 20% should be added to the material to cover consolidation and packing. If bought by weight, the following is the approximate amount required per cubic metre of compacted material:

Brick ballast   1,800 kg

Stone ballast   2,400 kg

On contracts where large volumes of hardcore are used, it is usual to have the material tipped and then spread, levelled and compacted by mechanical means. However, below are typical labour constants for spreading and levelling hardcore by hand. Compacting layers and surfaces are deemed to be included irrespective of depth and number of layers when using NRM2; however, although no longer required to be included in the description, these items still need to be considered, priced and included in the rate by the contractor/sub-contractor if required:

Item	Labourer hours
Imported filling as bed over 50 mm but not exceeding 500 mm deep, 350 mm finished thickness	$1.5/m^3$
Ditto exceeding 500 mm deep	$1.0/m^3$
Surface treatments compacting filling	$0.2/m^2$
Surface treatments trimming slopping	$0.40/m^2$
Blinding bed, not exceeding 50 mm thick, level to falls, cross falls or cambers, 40 mm finished thickness	$0.50/m^3$

Example

Imported filling as bed over 50 mm but not exceeding 500 mm deep, average 350 mm finished thickness. Cost per m^3.

A medium-sized mechanical shovel will spread and level approximately 15 m^3 of hardcore per hour; it is then consolidated in 150 mm layers by a roller.

Data

Hardcore £30 per m^3 delivered to site
Mechanical shovel including driver and fuel £25 per hour
Roller including fuel and driver £20 per hour

1 m^3 hardcore @ £30 per m^3 delivered to site and tipped	30.00	
Add 20% consolidation	6.00	
Hardcore per m^3		36.00

Laying and consolidating per hour	
Mechanical shovel	25.00
Roller	20.00
Labour	15.75
	£60.75

$$\text{Per m}^3 \frac{£60.75}{15\text{m}^3} \qquad \underline{4.05}$$

£40.05

Add profit and overheads 15%   <u>£6.01</u>

Cost per m³   <u>£46.06</u>

EARTHWORK SUPPORT – SPECIFICALLY REQUESTED BY CONTRACT DOCUMENTS

NRM2 Clause 5.8 states: *'This work shall only be measured where it has been specifically specified in the contract documents or if the contractor has been instructed by the contract administrator to provide the support during the course of the works.'*

Although now a deemed to be included item in NRM2, the cost of safely upholding the sides of excavations has to be calculated. In some instances, it may be a specific requirement of NRM2. Earthwork support refers to everything required to support the sides of the excavation, including in some instances special shoring or sheet piling (measured in accordance with NRM2, Section 7). The type of support required including the size and disposition of the various members will depend largely upon the depth and nature of the excavation, the stability of the ground to be upheld, the proximity of adjoining buildings, and the vibration from any adjoining roads. When pricing earthwork support, the estimator must decide the following:

- the type of poling boards to be used;
- the type of struts, e.g. timber, metal, etc.; and
- the section of the timber members, the spacing of the poling boards, etc. and the number of each times each item can be used.

If the sides of the excavation are up to 4 m apart then it's possible to span across from the excavation with supports; if over 4 m then it's necessary to provide raking shores to either side as illustrated in Figure 4.4.

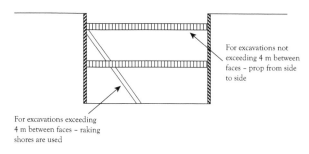

For excavations not exceeding 4 m between faces – prop from side to side

For excavations exceeding 4 m between faces – raking shores are used

Figure 4.4 Support to trench excavation

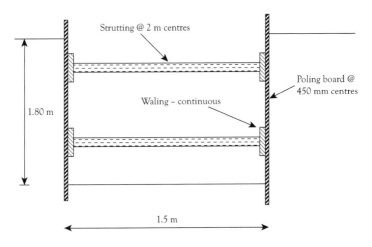

Figure 4.5 Conventional earthwork support for reasonably stable ground

Example

Earthwork support to foundation excavation not exceeding 2 m deep (see Figure 4.5).

Data

200 × 38 mm poling board £16.00/m
100 × 100 mm strutting £9.00/m
200 × 38 mm waling £16.00/m

Assume 10 uses
Assume 10 linear metres of trench = 36 m² of trench to be supported

Materials

	£
4 × 10 = 40 m 200 × 38 mm walings @ £16.00/m	640.00
200 × 38 mm poling boards @ 450 mm centres =	
5 × 1.80 m × 4 = 36 m @ £16.00/m	576.00
100 × 100 mm strutting @ 2 m centres =	
2 × 6 × 1.50 = 18 m @ £9.00/m	<u>162.00</u>
	1,216.00
Add waste 10%	<u>121.60</u>
	1,337.60

÷ 10 uses – cost per use	133.76

Labour

Fixing and stripping timber	
0.55 hours carpenter per m of trench =	
5.5 hours carpenter/m² @ £23.16 hour	127.38
Cost per 36 m² of trench	261.14
÷ 36 – cost per m	7.25
Add profit and overheads 15%	1.09
Cost per m²	£8.34

Other factors that could impact on the pricing of excavating and filling items are:

Work below ground-water level

Excavation below ground-water level is measured extra over (in addition to) any type of excavation, irrespective of depth. The factors influencing the estimator's judgement on excavation below ground-water level are as follows:

- operative working in the excavations may require protective clothing;
- outputs for both hand and machine excavation are considerably reduced; and
- the use of pumping should be considered to achieve optimal working conditions.

Services

NRM2 Clause 5.7 states: 'items are measured where there is a risk of the existing service being affected by the excavation process. The method of protection is left to the discretion of the contractor.'

Excavation adjacent to, across or under existing services is also measured extra over any type of excavation. The services may be such items as water mains, oil-pipe lines, gas pipes and electrical cables, as well as water pipes, telephone wires, district heating mains, culverts, sewers and drain pipes. The estimator must assess the impact on outputs based on:

- utility companies, such as gas and electricity, will need to be consulted, and the degree to which they will dictate the nature of how the work must proceed can vary;
- machine use will be restricted;
- hand excavation may be the only option; and
- damage to services may result not only in embarrassment but also financial penalties.

Additional working space

Although NRM2 has no specific requirement to measure working space when excavation has to be carried out in cramped or inaccessible situation, there may be occasions where the contractor or sub-contractor needs to allow for additional excavation and any associated disposal, filling and support in the pricing. Whether or not a working space allowance needs to be included will be apparent by reference to the relevant method statement.

SECTION 8: UNDERPINNING

Underpinning is a technique used for existing structures that have suffered differential settlement, caused by proximity to trees, landfill, mine workings, etc. It is carried out in small sections and involves excavating beneath the existing foundations and supporting them with a new substructure, wedging and pinning the new to the existing, and transferring the loads to a sound bearing level. The rates for underpinning will vary considerably depending on the ground conditions, the condition of the existing building and the perceived risk. Underpinning operations usually involve several trade such as excavation, concrete work, masonry, etc., although due to the possible restrictions placed on working conditions, the normal labour constants need to be adjusted to compensate for this. In some cases, hand digging may be necessary, and when the underpinning is extensive or complicated, it may be good practice to ask the contractors to submit a method statement with their bid (see Chapter 5) to ensure that all the risks have been identified and will be managed safely.

NRM2 allows underpinning to be measured and described in two ways:

- In the case of works that are 'not extensive', it can be included by means of a brief description of work, stating depth, maximum width and method of underpinning. Within this heading, items of foundations, walls and bases are kept separate. Therefore the onus is put very much onto the contractor/sub-contractor to design and price the works.
- In the case of works that are thought to be extensive (note there is no definition of 'extensive' given in NRM2), the work involved is measured in accordance with the appropriate NRM2 work sections.

Underpinning generally involves digging down to the underside of the existing foundations, then digging, in short lengths, underneath the existing foundation until firm bearing ground is found. Then new foundations and brickwork are built to the underside of the existing foundation. There are other underpinning systems available including mini piles and injection. The following are average labour constants for underpinning work:

Description	Unit	Labourer	Skilled operative
Excavate for preliminary trench down to existing foundation depth not exceeding 2 m	m³	6 hours	
Ditto over 2 m not exceeding 4 m	m³	9 hours	
Ditto over 4 m not exceeding 6 m	m³	11 hours	
Excavate below base of existing foundations	m³	8 hours	
Removing excavated materials	m³	4 hours	
Add for staging out per 1.5 m depth below ground level	m³	4 hours	
Concrete in foundations not exceeding 300 mm thick	m³	6 hours	
Ditto exceeding 300 mm thick	m³	5.50 hours	
One-brick wall in common bricks	m²	2.50 hours	2.50 hours
One-brick wall in engineering bricks	m²	3.75 hours	3.75 hours
Wedging and pinning to existing foundations	m²	2 hours	2 hours
Cutting away existing brickwork	m²	4 hours	4 hours
Formwork to face of concrete	m²	3.5 hours	

There follows two examples of the build-up of rates for some of the operations involved in underpinning:

Example

Excavation below the level of the existing foundation, maximum depth over 2 m not exceeding 4 m deep, commencing 3 m below existing ground level.

Labour

		£	
Excavate preliminary trench	8 hours		
Staging out 3 × 4 hours	12 hours		
Clearing away	4 hours		
	24 hours @ £15.75	£378.00 per m³	

Example

Wedging and pinning new brickwork to existing including dry packing.

£

Materials

0.3 m³ dry packing @ £25/m³   7.50

Labour

2 hours bricklayer @ £23.16 = £46.32
2 hours labourer @ £15.75 =   £31.50   <u>77.82</u>
                                                        <u>£85.32</u> per m²

SECTION 11: CONCRETE WORK

This section of the works includes:

- in-situ concrete work;
- formwork; and
- reinforcement.

In-situ concrete work

Factors to be taken into account when pricing concrete work items are:

- Whether ready-mixed concrete is to be used or if concrete is to be mixed on site. It is usually found that where access is not a problem and where reasonably large quantities of concrete are required over a regular period then it is better and more economical to set up a batching plant to mix on site. On a restricted site or where small quantities of concrete are required then it is probably better to use ready-mixed concrete, as it will be the cheaper alternative.
- Assuming that concrete is to be mixed on site, whether bagged or bulk cement is to be used. Bagged cement is delivered in 25 kg bags and has to be unloaded and stored in a dry location, whereas bulk cement is stored in a silo next to the mixer and is cheaper. Bagged cement also tends to be more wasteful.
- If concrete is to be mixed on site, the best position for the mixing plant from the point of view of transporting the mixed concrete around the site.
- The type and size of mixer to be used based on the output required; that's to say, how much mixed concrete can be produced per hour.
- The method of hoisting, placing and compacting the mixed concrete and whether a tower crane is to be used. Concrete may be transported by dumper, barrow or pumped.
- The cost of any measures necessary to protect the poured concrete either due to excessive drying out in hot weather or damage from frost and low temperatures.
- Common concrete mixes are described in the tender documents by their strength requirements, however this relates back to the volume of the ingredients. For example, plain in-situ concrete C20 (20 Newton/28 day strength) refers to a mix that is, by volume, one part Portland cement, two parts sand and four parts aggregate. When water is added during the mixing process, the dry materials combine together and reduce in volume by approximately 40%, although the percentage will depend on the mix of the concrete.

Quotations for sand and gravel can be either in tonnes or cubic metres; they can be adjusted as follows:

Material	Tonnes/m^3
Cement in bags	1.28
Cement in bulk	1.28–1.44
Sharp sand (fine aggregate)	1.60
Gravel (course aggregate)	1.40

On rare occasions, concrete may have to be mixed by hand and 4 hours/m^3 per labourer should be allowed for hand mixing concrete on a wooden board.

Labour constants

PLAIN IN-SITU CONCRETE

The following constants are average labour constants for mechanical mixing, transporting up to 25 m, placing and compacting:

Item	Hours/m^3 (labourer)
Concrete in trench filling	2.0
Horizontal work exceeding 300 mm thick in structures	1.5
Sloping work not exceeding 15° exceeding 300 mm thick in structures	3.0

REINFORCED IN-SITU CONCRETE

The following constants are average labour constants for mechanical mixing, transporting up to 25 m, placing and compacting, packing around reinforcement and into formwork if necessary:

Item	Hours/m^3 (labourer)
Concrete in trench filling	4.0
Sloping work not exceeding 15° exceeding 300 mm thick in structures	4.5
Vertical work exceeding 300 mm thick in structures (walls/slabs)	6.0
Vertical work exceeding 300 mm thick in structures (columns)	8.0

Example

Reinforced in-situ concrete (C25, 1:2:4), horizontal work not exceeding 300 mm thick in structures. Cost per m^3.

Data

Portland cement £190.00 per tonne delivered to site in bags on pallet

Sand £15.00 per tonne delivered to site

Course aggregate £14.00 per tonne delivered to site

Alternatively sand and aggregate may be delivered mixed as ballast.

Materials

	£	£
1 m³ cement = 1,400 kg cement @ £190.00 per tonne	266.00	
2 m³ sand = 3,200 kg @ £15.00 per tonne	48.00	
4 m³ aggregate = 5,600kg @ £14.00 per tonne	78.40	
	392.40	
Add shrinkage 40%	156.96	
	549.36	
Add waste 2.5%	13.73	
Cost per 7 m³	£563.09	
÷ 7 – cost per m³	£80.44	80.44

Mixing

Assume 200 litre mechanical fed mixer @ £40.00 per hour

Output 4 m³ per hour – cost per m³   10.00

Placing

4 hours labourer @ £15.75	63.00
	153.44
Add profit and overheads 15%	23.02
Cost per m³	£176.46

Mixer outputs in m3 of concrete per hour

Hand loaded:

100 litre	150 litre	175 litre	200 litre
1.2	1.8	2.1	2.4

Mechanical feed:

200 litre	300 litre	400 litre	500 litre
4.0	7.2	9.6	12.0

Transporting concrete

Once mixed concrete has to be transported from the batching plant to the point on the site where it is to be placed, this can be done in several ways:

Wheeling by barrow

When wheeling and depositing concrete by barrow, allow a constant of 1 hour per labourer per cubic metre per 100 m round trip.

Example

Using a 200 litre mixer with a 4½ minute mix and a round trip of 200 meters.

Mixer output = 2.4 m^3

Total labour $1.00 \times \dfrac{200}{100}$ metres x 2.4 m^3 = 4.8 hours general operative

Transporting by dumper

For transporting larger quantities of mixed concrete, a dumper can be used, although this should be restricted to distances of less than about 400 m otherwise separation of the concrete ingredients could take place. A 1 tonne dumper with a capacity of 6 m^3 should be able to cope with the output (4 m^3) of a 200 litre mechanically fed mixer. A dumper truck can travel at approximately 25 km per hour.

Assuming an hourly cost for a dumper of £30 per hour including consumables plus driver costs of £15.75 per hour plus £0.96 for additional skills (NRW) = £46.71 per hour.

Cost of using dumper $\dfrac{£46.71}{4 \text{ m}^3}$ = £11.68 per m^3

Hoisting

Take the full cost per hour of the hoist man during concreting operations.

Pumping concrete

Pumping is a very efficient and reliable means of placing concrete; boom pumps and line pumps are available. Sometimes a pump is the only way of placing concrete in certain locations, such as high-rise buildings or large areas of slabs. The speed of concrete delivery when pumping is used should be considered when calculating the number of labourers required to place/tamp/vibrate the concrete when in the form-work. Most standard concrete mixes can be pumped with little or no modification; however, several factors affect the pumpability of the concrete and job factors such as aggregate, conveying line size and pumping equipment must be considered when designing a concrete mix for pumping. A quotation will be required from a company specialising in concrete pumping.

Formwork

Timber formwork

Prior to commencing pricing this section, the estimator, possibly in conjunction with the planning department, should examine the tender drawings in order to decide the design of the formwork and the number of uses that can be obtained. These can be categorised as follows and depend on the nature of the work to be sup-ported at its position in the works:

- for large uninterrupted areas – up to 10 uses;
- medium complex areas – 3–6 uses; or
- more complex areas – 3 uses.

Timber formwork is formed from plywood shuttering supported by props (Figure 4.6). It should be noted that when reusing formwork, an allowance should be made for cleaning, de-nailing and preparing the surfaces of the timber before reusing.

Other factors that should be considered when pricing formwork are:

- whether standard (as struck) or special finishes are required; and
- whether the formwork has to be left in place (in which case, only one use can be obtained).

A typical build-up when using timber formwork is as follows:

- cost of plywood shuttering, including waste, divided by the number of uses as above;
- bolts and nails and labour in the treatment/cleaning of the shuttering face;
- cost of metal propping; and
- labour fixing and striking the formwork, usually carried out by carpenters and labourers.

Metal/proprietary formwork

Among the advantages of steel formwork are the following:

- very strong and able to carry heavy loads;
- ease of fixing;
- uniform size and surface; and
- long life.

Among disadvantages of steel formwork are:

- limited size or shapes;
- excessive loss of heat; and
- very smooth finished surface, which may give problems for finishing process.

Glass-reinforced plastic formwork

Among the advantages of glass-reinforced plastic formwork are as follow:

- useful for complex shape and special features;
- easy to strike;
- lightweight; and
- damage to formwork can be easily repaired.

Among the disadvantages of glass reinforced plastic formwork is:

- comparatively high initial cost.

Example

Plain formwork, sides of foundations and bases 500 mm high. Cost per m.

Data

19 mm plywood £14/m^2
50 × 50 mm sawn timber for support/framing £3.00/m

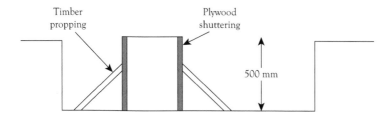

Figure 4.6 Example of timber formwork to sides of foundations

	£
Assume 10 m²	
10 m² 19 mm plywood @ £14.00/m²	140.00
Add waste 10%	14.00
50 × 50 mm sawn timber support calculated	
from drawings – 12m @ £3.00/m	36.00
Add waste 10%	3.60
Labour making 10 hours carpenter @ £23.16	231.60
	425.20
Assume 6 uses ÷ 6 – cost per use	70.87
1 kg bolts and nails	2.50

Labour

Fixing 15 hours carpenter @ £23.16	347.40
Strip and clean	
8 hours carpenter @ £23.16	185.28
4 hours labourer @ £15.75	63.00
Cost per 10 m²	669.05
÷ 10 – cost per m²	66.91
Cost per m – 500 mm wide	33.46
Add profit and overheads 15%	5.02
	£38.48

Example of timber formwork to soffits of suspended slab 250 mm tick

Plain formwork soffits of horizontal work for concrete not exceeding 300 mm thick, propping over 3 m not exceeding 4.5 m high. Cost per m².

Data

19 mm plywood £14/m² delivered in 2.4 × 1.2 m sheets
Props £10.00 per week

From the engineer's drawings, it is established that each bay of shuttering is 4 m × 3 m. Delivered in 2.4 x 1.2 m sheets, the plywood will be supported with telescopic props at 400 mm centres.

Materials

Based on a 3 × 4 m bay = 12 m²	
12 m² 19 mm plywood @ £14/m²	168.00
Waste 7.5%	12.60
	180.60
Assume 10 uses ÷ 10	18.06

Sundries
Mould oil
Nails 1.00
Telescopic props
Required @ 400 mm centres spanning 4 m =
10 + 1 = 11 props @ £5 per week for 3 weeks .... <u>165.00</u>
Cost per 12 m² .... <u>184.06</u>
÷ 12 – cost per m² 15.34
Labour per m²
Fixing 1.30 hours carpenter @ £23.16 .... 30.11
Strip and clean
0.5 hours labourer @ £15.75 .... <u>7.88</u>
  53.33
Add profit and overheads 15% .... <u>8.00</u>
Cost per m² .... <u>£61.33</u>

Reinforcement

It is usual for bar reinforcement to be delivered to site, cut to length and bent in accordance with the bending schedules. Steel fixers are entitled to additional payments in accordance with the Working Rule Agreement. Reinforcement is usually fixed by black tying wire. For binding wire and rolling margin, 5% should be added. Allow four hours per tonne labourer for unloading.

Labour constants

Labour cutting, bending and fixing reinforcement, per 50 kg:

Item	Hours
10 mm diameter bars	4.00
10 mm–16 mm bars	3.00
Over 16 mm bars	2.75

Figure 4.7 Plastic bar reinforcement chair

Figure 4.8 Plastic circular reinforcement spacer

Fabric reinforcement

As an average 30 m² per hour should be allowed for cutting and fixing fabric reinforcement. An allowance of 12.5%–15% for waste and laps should be made, and the areas in the bills of quantities will be net areas and exclude laps between sheets.

Example

Mesh reinforcement Ref. A252 weighing 3.95 kg/m² with 150 mm minimum side and end laps. Cost per m².

Data

A252 mesh reinforcement, weighing 3.95 kg/m² size 4,800 mm × 2,400 mm – £74.12 (11.52 m²)
Assume 23 m² (approximately two sheets)

Materials

	£
23 m² of fabric A252 reinforcement delivered to site	148.24
Add laps and waste 15%	<u>22.24</u>
	170.48
Labour $\dfrac{23}{30}$ hours steelfixer @ £23.16	<u>17.76</u>
Cost per 23 m²	188.24
Add profit and overheads 15%	<u>28.24</u>

	216.48
÷ 23 – cost per m²	£9.41

Example

12 mm diameter straight mild steel bars – Cost per tonne

Data

12 mm diameter mild steel bars delivered to site £305.00 per tonne

Materials	£
1 tonne 12 mm diameter mild steel bar	305.00
Add waste 1%	3.05
	308.05
Tying wire: 10 kg per tonne @ £19.80/kg	19.80
Chairs and spacers, say	12.00
	339.85

Labour	
40 hours steel fixer @ £23.16	926.40
	1,266.25
Add profit and overheads 15%	189.94
Cost per tonne	1,456.19
÷ 1000 – cost per kg	£1.46

Precast concrete units

Items such as lintels, sills, copings and plank floors come to site as precast units ready to be placed into position. To the cost of the precast units, the additional costs of the following items should be considered:

- unloading and stacking;
- hoisting and bedding in place;
- material for bedding;
- waste (damaged or broken units) – 2.5%; and
- profit and overheads.

Allow 0.30 hours bricklayer and labourer per linear metre for hoisting and fixing a 225 × 150 mm lintel up to 3 m above ground.

SUNDRY LABOUR CONSTANTS

Description	Unit	Hours (bricklayer and labourer)
150 × 75 mm sill	no.	0.20
300 × 100 mm coping	m	0.35
115 × 150 mm lintel	no.	0.15
225 × 225 mm lintel	no.	0.35
300 × 225 × 150 mm padstone	no.	0.50
600 × 600 × 100 mm pier cap	no.	1.00

Example

300 × 100 mm precast concrete (mix B) weathered and twice-throated coping bedded in cement mortar (1:3). Cost per no.

Data

305 × 600 mm precast concrete coping £13.08 each
Cement mortar (1:3) £116.91/m³ (taken from masonry section)

Materials

Assume 3 m	
10 no. 305 × 600 mm precast concrete coping	£130.80
0.004 m³ mortar/m @ £116.91/m³ =	
0.012 m³ @ £116.91/m³	1.40
	132.20
Add waste 2.5%	3.31
	135.51

Labour

0.45 hours @ £23.16	
£15.75	
£38.91 per hour	17.51
	153.02
Add profit and overheads 15%	22.95
Cost per 3 m	175.97
÷ 3 – cost per m	£58.66

Example

400 × 400 mm precast (mix B) reinforced precast concrete pier cap bedded in cement mortar (1:3) Cost per no.

Data

400 × 400 mm pier cap £13.14
Cement mortar (1:3) £116.91/m³ (taken from masonry section)

Materials

400 × 400 mm pier cap £13.14	13.14
0.002 m³ mortar @ £116.91	<u>0.23</u>
	13.37
Add waste 2.5%	<u>0.33</u>
	13.70

Labour

0.20 hours @ £23.16	
£15.75	
£38.91 per hour	<u>7.78</u>
	21.48
Add profit and overheads 15%	<u>3.22</u>
Cost per no.	<u>£24.70</u>

Example

115 × 150 × 1500 mm precast concrete (mix A) lintel, reinforced with and including two No 10 mm diameter mild steel bars. Cost per m.

Data

115 × 150 × 1500 mm precast concrete lintel £39.79 each
Cement mortar (1:3) £116.92/m³ (taken from masonry section)

Materials

115 × 150 × 1500 mm precast lintel	39.79
0.002 m³ mortar @ £116.91	<u>0.23</u>
	40.02
Add waste 2.5%	<u>1.00</u>
	41.02

Labour

0.15 hours @ £23.16	
£15.75	
£38.91 per hour	<u>5.84</u>

	46.86
Add profit and overheads 15%	7.03
Cost per no.	£53.89

SECTION 14: MASONRY

Brick/block walling

Mortar

When estimating the cost of masonry, in addition to the bricks or blocks the cost of the mortar also has to be calculated. Mortar comes in a variety of mixes depending on location and type of bricks or blocks being used; mortar is usually made from cement and sand. Generally, the mortar should be weaker than the bricks or blocks, so if pressure is placed on a brick, it is the mortar that should fail without causing the brickwork to crack. NRM2 requires that the type of pointing should be stated and this is particularly relevant to facing brickwork; pointed brickwork requires additional labour and therefore cost. Common types of pointing are weather struck (or struck), flush and bucket-handle joints.

Example

Cement mortar (1:3). Cost per m³.

Materials

	£	£
1 m³ cement = 1,400 kg cement @ £190 per tonne	266.00	
3 m³ sand = 4,800 kg @ £15.00 per tonne	72.00	
	338.00	
Add shrinkage 25%	84.50	
	422.50	
Add waste 5%	21.13	
Cost per 4 m³	£443.63	
÷ 4 – cost per m³	£110.91	110.91

Mixing

Assume 100 litre mixer @ £24.00 per hour	
Output 4 m³ per hour – cost per m³	6.00
Cost per m³	£116.91

For small quantities of mortar mixed by hand on a board, allow 4 hours/m³ labourer.

The amounts of mortar required will vary according to the size of the bricks as follows (Figure 4.9):

	55 mm brick	65 mm brick	75 mm brick
Per 102.5 mm thickness of wall	0.035	0.03	0.028
Per vertical joint (perpend)	0.01	0.01	0.01

The amount mortar required per m² of half-brick wall using 65mm bricks is:

Bed joint	0.03 m³
2 × perpends	0.02 m³
	0.05 m³

Bricks

The usual size for a clay brick is 215 × 102.5 × 65mm with a 10 mm mortar joint making the nominal size 225 × 112.5 × 75 mm. A 112.5 mm thick wall is referred to as a half-brick wall and there are 60 bricks per m² in a half-brick wall laid stretcher bond. Once upon a time, bricks were delivered to the site loose in a lorry and tipped, causing a great deal of waste; now bricks are packed on pallets in polythene and unloaded by mechanical hoist, thereby cutting down the amount of wastage and reducing the need for stacking. The amount of wastage, however, will vary considerably and depends on the nature and complexity of the work; allow between 7.5 and 12.5% for cutting and waste.

For work in facings, the number of bricks per m² will vary according the specified brick bond and the thickness of the wall as follows:

Stretcher bond   60 per half-brick wall

Header bond   120 per half-brick wall

English bond   90 per one-brick wall

Flemish bond   80 per one-brick wall

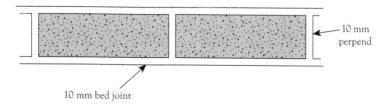

10 mm perpend

10 mm bed joint

Figure 4.9 Joints in brickwork

LABOUR CONSTANTS

The productivity of labour will depend again upon the nature of the work and, in addition, the organisation of the gang. A bricklaying gang is made up from bricklayers and labourers. It is the responsibility of the labourers to keep the bricklayers supplied with bricks and mortar so that they are able to maximise their output without having to keep breaking off to replenish their materials. It is usual to have two bricklayers to one labourer.

Example

Walls one brick thick in common bricks laid English bond in cement mortar (1:3) – Cost per m^2.

Data

Common bricks £500.00 per 1,000 delivered to site

Assume 10 m^2

Materials

	£	£
900 common bricks @ £500.00 per 1,000	450.00	
Add waste 7.5%	33.75	

Mortar

	£	£
10 m^2 × 0.08 = 0.8 m^3 @ £116.91	93.53	
Add waste 5%	4.68	
	£581.96	581.96

Labour

Based on gang rate:

2 bricklayers and 1 labourer

$$2 × £23.16 = £46.32$$
$$£15.75$$

Hourly rate per gang   £62.07

1 bricklayer can lay 60 bricks per hour, therefore
output = 120 bricks per gang hour ÷ $\frac{90}{120}$ m^2 = £46.55/m^2

10 m^2 @ £46.55   465.50

Cost per 10 m^2	1047.46
Add profit and overheads 15%	157.12
	1204.58
÷ 10 – cost per m^2	£120.46

Example

Walls half-brick thick in LBC Tudor facings laid stretcher bond in cement mortar (1:3) with weather-struck joint one side. Cost per m^2.

Data

Tudor facings: £1,000.00 per 1,000 delivered to site

Assume 10 m^2

Materials

Facings:

	£	£
600 facings @ £1,000.00 per 1,000	600.00	
Add waste 7.5%	45.00	

Mortar:

	£	£
10 m^2 × 0.05 = 0.5 m^3 @ £116.91/m^3	58.46	
Add waste 5%	2.92	
	706.38	706.38

Labour

2 bricklayers and 1 labourer

$$2 × £23.16 = £46.32$$
$$£15.75$$

Hourly rate per gang   £62.07

1 bricklayer can lay 60 bricks per hour, therefore
output = 120 bricks per gang hour = 2 m^2 = £31.04/m^2

10 m^2 @ £31.04/m^2	310.40
Cost per 10 m^2	1,016.78
÷ 10 – cost per m^2	101.68
Add profit and overheads 15%	15.25
Cost per m^2	£116.93

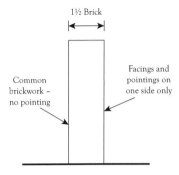

Figure 4.10

Example

This example is based on a wall one and half bricks thick, built in common brick-work but with facings on one side to form a fair exposed face (Figure 4.10).

One and half wall in common bricks in cement mortar (1:3) one side finished in LBC Tudor facings (prime cost £1,000.00 per 1,000 delivered to site) built in Flemish bond and pointed one side with weather-struck joint. Cost per m².

Assume 10m²

Data

	£	£
Common bricks £500.00 per 1,000 delivered to site		
LBC Tudor facings £1,000.00 per 1,000 delivered to site		

Materials

Number of common bricks/m² = 60 × 3 = 180		
Number of facings/m² =	<u>80</u> × 10 = 800	
Net number of common bricks	100 × 10 = 1,000	
800 facings @ £1,000.00 per 1,000		800.00
1,000 common bricks @ £500.00 per 1,000		<u>500.00</u>
		1,300.00
Add waste 7.5%		<u>97.50</u>
		1,397.50
Mortar:		
10 × 0.15 m³ mortar = 1.50 m³ @ £116.91/m³		175.37
Add waste 5%		<u>8.77</u>
		1,581.64

Labour

Output = 120 bricks per gang hour ÷ 2 m² = £31.04/m²

$$\frac{1800}{120} = 15 \text{ gang hours @ £31.04} \qquad \underline{465.60}$$

	2,047.24
Add profit and overheads 15%	307.09
Cost per 10 m²	2,354.33
÷ 10 – cost per m²	£235.43

Sundry labour constants – hours per m²

Item	Bricklayer	Labourer
Forming cavity 50 mm wide including building in wall ties at 5/m²	0.10	0.05
Damp-proof course not exceeding 300 mm wide	0.20	0.10

Extra over

Some masonry items are required to be described as extra over and the estimator must calculate the extra cost of the labour and material involved with these items.

Sundry brick/block walling labour constants – hours per m

Item		
Extra over walls for closing cavities with brickwork at opening perimeters	0.35	0.18
NRM2 deemed to be included items		
Bonding half-brick wall to existing	1.10	0.55
Bedding and pointing frames	0.05	0.025

Work in existing buildings

Sundry labour constants – hours per no.

Item		
Holes for pipes 50 mm diameter in half-brick wall	0.25	0.15
Ditto over 110 mm diameter in half-brick wall	0.30	0.25
Ditto over 150 mm diameter in half-brick wall	0.55	0.35

Example

One-layer Visqueen polyethene damp-proof course not exceeding 300 mm wide horizontal bedded in gauged mortar (1:1:6). Cost per m.

Data

Visqueen damp-proof course 112.5 mm wide, £20 per 20 m delivered to site
Assume 10m

£

Materials

10 m Visqueen damp-proof course @ £20.00/20 m	10.00
Add waste and laps 5%	0.50

Labour

0.10 hours bricklayer @ £23.16	£2.32	
0.05 hours labourer @ £15.75	£0.79	
	£3.11 per m	
10 m @ £3.11 per m		31.10
		41.60
Add profit and overheads 15%		6.24
Cost per 10 m		47.84
÷ 10 – cost per m		£4.78

Note that it is not necessary to include the cost of mortar for bedding the damp-proof course with this item as this has previously been included with the masonry price.

Blockwork

Blocks are heavier and larger than bricks; the usual size is 440 × 215 × 100 mm thick, excluding the mortar joint. There are three main types of block: aerated, dense and hollow clay. Like bricks, blocks are delivered to site stacked on pallets for easy unloading and use. Large than bricks, blocks need more time for transporting around the site and hoisting in place. Allow 5% waste.

Example

Walls in lightweight concrete blocks (7.3N) in gauge mortar (1:1:6). Cost per m^2. Assume 10m^2

Data

$440 \times 215 \times 100$ mm aerated concrete blocks (7.3N) £13.00 per m² delivered to site
Gauge mortar; hydrated lime 0.6 tonnes/m³ £25.00 per tonne
Gauge mortar (1:1:6)

Materials

	£	£
1 m³ cement = 1,400 kg cement @ £190.00 per tonne	266.00	
1 m³ hydrated lime = 600 kg @ £30.00 per tonne	18.00	
6 m³ sand = 9,600 kg @ £15.00 per tonne	144.00	
	428.00	
Add shrinkage 25%	107.00	
	535.00	
Add waste 5%	26.75	
Cost per 8 m³	561.75	
÷ 8 – cost per m³	£70.22	70.22

Mixing

Assume 100 litre mixer @ £20.00 per hour	
Output 4 m³ per hour – cost per m³	5.00
Cost per m³	£75.22

Blocks

10 m² blockwork @ £13.00/m²	130.00
Waste 10%	13.00
Mortar 0.01 m³ per m² = 0.1 m³ @ £75.22	7.52
Add waste 5%	0.38
	150.90

Labour

0.50/m² per gang per m² = 5 gang hours @ £62.07	310.35
	461.25
Add profit and overheads 15%	69.19
Cost per 10 m²	530.44
÷ 10 – cost per m²	£53.04

SECTION 16: CARPENTRY

Timber framing

Timber first fixings

This section includes such items as: rafters and associated roof timbers, wall plates, roof and floor joists, beams, posts or columns, partition and wall members, and strutting. Of course, in today's industry most roof structures are in the form of pre-fabricated roof trusses, but it still may be necessary estimate the rate of individual roof timbers, particularly in the case of flat roofs, which are usually formed in-situ.

LABOUR CONSTANTS

Note: the following constants relate a 50 × 100 mm timber; for larger or smaller sizes the rates should be adjusted accordingly:

Item	Hours (carpenter per m)
Wall plates	0.10
Floor joists	0.15
Partitions	0.30
Ceiling joists	0.15
Flat-roof joist	0.17
Strutting	0.30

There are two general groups of wood – hardwood and softwood – and these are defined by a number of characteristics. Hardwood is timber produced from broad-leaved trees and soft wood from coniferous trees. Structural timbers, such as floor and roof joists, are stress graded according to the proportion of defects. For softwood, the gradings are divided into GS (General Structural) and SS (Special Structural). Timber is further defined by species, e.g. Scots pine. Hardwoods have only one grad-ing. Sizes for timber sections are stated in two ways: ex and finished. Unplaned or sawn timber is referred to as 'ex'; once planed, it becomes wrot timber and is referred to as finished. Therefore, timber sections may be referred to in two ways depending on whether the sizes quoted are 'ex' or finished:

• 50 × 25 mm ex softwood batten; or
• 46 × 21 mm finished softwood batten.

The finishing process planes 2 mm from each face. Softwood is sold by the linear, square and cubic metre.

For measurement purposes, NRM2 divides timber into the following: carpentry/carcassing; structural/timber framing such as roof and floor joists, roof trusses, etc.

Timber in this category is generally sawn softwood, treated to protect it against attack by decay and rot. Most modern roofs are constructed from roof trusses or trussed rafters that are prefabricated and delivered to site ready for lifting into position by crane. The timbers, which are typically 80 × 40 mm, are held together with special metal plates. Using roof trusses has several advantages over traditionally in-situ roof construction, namely:

- speed;
- skilled labour is not required;
- spans of up to 12 m can be accommodated; and
- cost savings compared to traditional construction.

Once hoisted into position, it's necessary to secure the roof trusses with softwood binders and galvanised metal straps to prevent distortion.

CARPENTRY/CARCASSING

Example

47 × 150 mm softwood floor joists. Cost per m.

Data

47 × 150 mm treated softwood £16.50/per 3 m delivered
Assume 9 linear metres

Materials

	£
9 m 47 × 150 mm softwood joists @ £16.50/3 m	49.50
Add waste 7.5%	3.71
	53.21

Labour

10 × 0.1 = 1 hour carpenter @ £23.16	23.16
	76.37
Add profit and overheads 15%	11.46
	87.83
÷ 9 – cost per m	£9.76

Example

47 × 50 mm softwood strutting. Cost per m.

Data

47 × 50 mm treated softwood £8.20 per 3 m delivered
Assume 9 linear metres

Materials

	£
9 m 47 × 50 mm softwood joists @ £8.20 per 3 m	24.60
Add nails 0.1 kg per m = 1kg @ £3.00/kg	3.00
	27.60
Add waste 7.5%	2.07
	29.67

Labour

10 × 0.1 = 1 hour carpenter @ £23.16	23.16
	52.83
Add profit and overheads 15%	7.93
	60.76
÷ 9 – cost per m	£6.75

Example

100 × 75 mm softwood treated wall plate. Cost per m.

Data

100 × 75 mm softwood treated wall plate £21.57 per 3 m delivered to site
Assume 9 linear meters

Materials

9 m 100 × 75 mm sawn softwood @ £21.57 per 3 m	64.71
Add waste 7.5%	4.85
	69.56

Labour

Hours carpenter per 1 m

1 hour carpenter @ £23.16	<u>23.16</u>
	92.72
Add profit and overheads 15%	<u>13.91</u>
Cost per 9 m	<u>106.63</u>
÷ 9 – cost per m	<u>£11.85</u>

TIMBER, METAL AND PLASTIC BOARDING, SHEETING, DECKING, CASINGS AND LININGS, METAL AND PLASTIC ACCESSORIES

Example

18 mm flooring-grade horizontal chipboard flooring exceeding 600 mm wide. Cost per m².

Data

£15.30 per 2400 × 600 mm sheet delivered to site
Assume 10 m²

Materials

10 m² 18 mm chipboard flooring @ £15.30 per 1.44 m² sheet

$$\frac{10.00}{1.44} \times £15.30 = \qquad £106.25$$

Nails 0.1 kg/m² 1 kg @ £2.20/kg	<u>£2.20</u>
	£108.45
Add waste 10%	<u>£10.84</u>
	£119.29

Labour

0.50 hours carpenter/m²	
5 hours @ £23.16	<u>£115.80</u>
	£235.09
Add profit and overheads 15%	<u>£35.26</u>
Cost per 10 m²	<u>£270.35</u>
÷10 – cost per m	<u>£27.04</u>

SUNDRY LABOUR CONSTANTS

Description	Unit	Labour (hours)
19 mm tongue and groove boarding to soffit	m	1.40
19 mm plywood soffit	m	0.20
25 mm plywood fascia	m	0.30
32 mm barge board – 150 mm wide	m	0.20

SECTION 17: SHEET ROOF COVERINGS

BITUMINOUS FELT

The most common form of material for covering flats roofs is felt roofing. Traditionally made from bitumen, there are now more high-tech, high-performance membranes available, based on polyester or similar material, that do not suffer from the main disadvantage of traditional materials, namely deterioration in 15–20 years. In the case of traditional bituminous felts, the material is supplied in rolls in a variety of weights and is laid usually in two or three layers. Felt is measured net but laid with side and end laps of 150 mm and these must be taken into account in the calculation. The finished roofing must be protected against solar radiation; the most common approach being with the application of chippings. The traditional method of applying bitumen felt is to torch it on but increasing this is being replaced by the cold process known as pour and roll. Allow 15% on net quantities to allow for laps and waste.

Example

Three-layer Chestermeric polyester-cored bitumen high performance pour and roll sloping roof coverings to falls not exceeding 10˚ from horizontal over 500 mm wide comprising: Chesterplus venting layer laid on and including 50 mm Rockwool rigid insulation, Chesterplus 250 sanded base layer and Chesterplus 250 green cap sheet finished with white spar chippings. Cost per m^2.

Data

Venting layer £40.00 per 8 × 1 m roll
250 Sanding layer £60.00 per 8 × 1 m roll
259 Green cap sheet £70.00 per 8 × 1 m roll
Easy-melt 105/35 bitumen £50.00 per 25 litres
50 mm Rockwool rigid insulation £6.00/m^2
Assume 10 m^2

Materials

10 m² venting layer @ £40.00 per 8 × 1 m roll	50.00
10 m² 250 sanding layer @ £60.00 per 8 × 1 m² roll	75.00
10 m² 250 green cap sheet @ £70.00 per 8 x 1 m² roll	<u>87.50</u>
	212.50
Add laps and waste 15%	31.88
10 m² of Rockwool insulation @ £6.00/m²	60.00
10 m² Spar chipping @ £6.00 per 20 kg	6.00
Easy-melt bitumen	
10 m² bitumen @ 2 litres per m² = 20 litres per layer	
60 litres + chipping 10 litres = 70 litres @ £50/25 litres	

$$\frac{70}{25} = 2.8 \times £50 = \qquad\qquad 140.00$$

Add waste 10%	<u>14.00</u>
	464.38

Labour

0.6 gang hours per m² roofer and labourer	
6 hours @ £23.16	
<u>£15.75</u>	
£38.91	<u>233.46</u>
	697.84
Add profit and overheads 15%	<u>104.68</u>
Cost per 10 m²	<u>802.52</u>
÷ 10 – cost per m²	<u>£80.25</u>

SUNDRY LABOUR CONSTANTS

Item	Gang hours per m²
One-layer felt nailed to timber	0.06
Application of primer to concrete surfaces	0.05
50 mm insulation board	0.10

	Gang hours per m
Boundary work – abutments 150 mm high	0.12
Ditto – 225 mm high	0.18
Eaves – 100 mm turn-down	0.18
Ditto – 150 mm ditto	0.22

Linings to gutters, etc. increase labour constants as follows:

Item	% increase
Width not exceeding 100 mm	500
Ditto 100–200 mm	200
Ditto over 300 mm	50

Flashings are an important part of flat roof coverings. The materials traditionally used for flashings are sheet metals such as lead and zinc, and the flashings, as can be seen in Figure 4.11, cover the joint between the membrane (felt roofing) upstand and the structure to stop water penetrating into the structure. Flashings are turned into a grove that is either cut into the structure in the case of concrete, or into the raked out brickwork joint in the case of masonry, and then pointed in cement mortar to complete the detail. Although in accordance with NRM2, raking out joints, etc., is deemed to be included, nevertheless the cost does have to be included by the estimator.

SECTION 18: TILE AND SLATE ROOF AND WALL COVERINGS

Pitched-roof coverings

The most common form of sloping roof coverings are slates and tiles; they are both fixed on softwood battens with nails, which are, in turn, laid on underfelt.

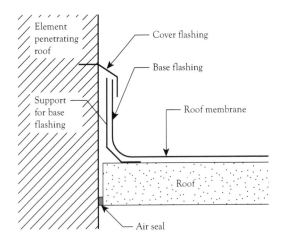

Figure 4.11 Flashing detail at abutment

In Scotland, in addition, sarking (treated sawn softwood boarding) is fixed directly to the rafters prior to the application of underfelt and battens. This practice is probably due to the harsher climate experienced north of the border. Roof coverings are now generally delivered to site on pallets, pre-packed and are usually ordered by the thousand.

Because the size of the slate or tile and the method of fixing affects the number of units required, it is necessary to calculate the number of slates or tiles per m^2, although this type of information is readily available in manufacture's literature that can be accessed online.

Slate roofing

Slates come in a variety of sizes with two nail holes and can be either head nailed or centre nailed. Occasionally, roofers have to form the fixing holes in slates on site using a special tool known as a zax. Head nailing tends to give better protection from the weather for the nails. A special shorter tile is used for the under-eaves course; an undercloak course is required at verges in order to divert water back onto the roof.

Therefore, to calculate the number of centre nail slates size 450 mm × 300 mm, laid with 100 mm lap, apply the following:

$$\frac{\text{length of slate} - \text{lap}}{2} \times \text{width of slate} \quad \frac{0.450 - 0.100}{2} \times 0.300 = 0.0525$$

$$\frac{1\text{m}^2}{0.0525} = 19.05 \text{ say 19 slates per m}^2$$

The number of head-nailed slates per square metre is calculated using the following:

$$\frac{\text{length of slate} - \text{lap} - 25\text{mm}}{2} \times \text{width of slate}$$

Note: non-ferrous nails such as aluminium or copper should be used for fixing as this reduces nail fatigue.

Slates for roofing come in a variety of sizes, varying from 650 × 400mm to 250 × 150 mm and very often are laid with diminishing courses towards the eaves. For this reason, the number of slates required has to be carefully calculated once the details are known.

Synthetic slate

Synthetic slates closely resemble their natural counterparts, but are lighter and much less expensive; the most common size is 600 mm × 300 mm and they are fixed as for centre-nailed natural slates, with a clip being used on the bottom edge to prevent lifting. They are fixed with copper nails. Also available is a synthetic slate made from a mixture of resin and crushed natural slate. These are single-lap interlocking slates and are fixed accordingly.

Tile roofing

Plain tiles

Plain tiles are known, like slates, as double-lap coverings to prevent water from penetrating the coverings from run off or capillary action. They are referred to as double lap because for part of their area the slate or tile laps two others in the course below. The most common size for a plain tile is 265 × 165 mm.

Interlocking tiles

All interlocking tiles are single lap, which means that there is only one layer of tiles on the roof, apart from the overlaps. Interlocking tiles are not, with the exception of the perimeter tiles, nailed.

Battens

To calculate the linear quantity of batten per square metre, divide the gauge of the tiles into 1 m (Figure 4.12).

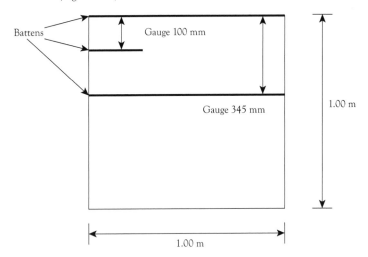

Figure 4.12 Calculation of quantities of battens

Therefore for 100 mm gauge: $\dfrac{1,000}{100} \times 1\,\text{m} = 10$ m per m^2 of roof.

For 345 mm gauge: $\dfrac{1,000}{345} \times 1\,\text{m} = 2.9$ m per m^2 of roof.

There is a number of free calculators available on manufacturer's web sites that calculate the quantity of tiles needed.

Nails for tiling

Approximate number of 44 mm nails per kilogram:

* copper: 290;
* composition: 280.

Note: tiles are usually nailed every fourth course.

Nails for battens

Approximate number of nails per kilogram:

* 50 mm × 10 gauge: 260;
* 63 mm × 10 gauge: 220.

Allow 5% waste on nails.

Labour constants and materials for plain tiling per m²

Lap of tile	Number of tiles	Tile nails (kg)	Battens (linear m)	Batten nails (kg)	Labour (tiler and labourer)
60	63	0.09	10.08	0.11	0.67
75	68	0.10	10.88	0.12	0.73
90	74	0.11	11.84	0.13	0.79
100	78	0.12	12.48	0.14	0.83

Fixing tile battens per 100 m 2.50 hours roofer/1.25 hours labourer, including unloading.

Example

Roof coverings 265 × 165 mm Marley Eternit Hawkins Staffordshire Blue plain clay roofing tiles with 75 mm head lap 40° pitch each tile nailed with 2 no. 44 mm composition nails to 38 × 25 mm treated sawn softwood battens at 100 mm centres to one layer of felt to BS Type 1F with 100 mm minimum laps fixed with galvanised clout nails. Cost per m².

Materials delivered to site

Hawkins Staffs Blue tiles £700.00 per 1,000 delivered to site
Underfelt   £13.00 per 15 m × 1 m roll
Nails for tiles   £2.00 per kg
Battens for tiles   £3.20 per 10 m
Nails for battens   £1.00 per kg
Assume 10 m²

Materials

	£	£
Tiles (from Marley web site) 68 per m²:		
680 Hawkins Staffs Blue tiles @ £700 per 1,000	476.00	
Add waste 10%	47.60	
Battens:		
100 m batten (see calculation above) @ £3.20 per 10 m	32.00	
Add waste 10%	3.20	
Nails for battens:		
1.2 kg @ £1.00 per kg	1.20	
Add waste 5%	0.06	
Nails for tiles:		
680 tiles × 2 = 1,360 nails @ 280/kg =		
4.86 kg nails @ £2.00	9.72	
Add waste 5%	0.49	

Underfelt:

15×1 m roll $= 15$ m²

$$\text{Cost per 10 m}^2 = \frac{10.00}{15.00} \times £\,13.00 = \qquad 8.67$$

	£	
Add laps 10%	0.87	
Add waste 5%	0.44	
		580.25

Labour

Tiles:

| 7.30 hours roofer and labourer | £23.16 | |
| | £15.75 @ £38.91 | 284.04 |

Tile battens:

2.5 hours tiler	23.16	57.90
1.25 hours labourer	15.75	19.69
		£941.88
Add profit and overheads 15%		£141.28
		£1,083.16
÷ 10 – cost per m²		£108.32

Sundry labour constants

Note: NRM2 defines boundary work as work associated with closing off or finishing off tile or slate roofing at the external perimeter, at the abutment with different materials or the perimeter of openings and voids. This typically means work at eaves, ridges and verges of roofs.

Description	Unit	Labour (hours)
Boundary work:		
Double course of tiling at eaves	m	0.05
Verges	m	0.05
Cutting to valleys	m	0.50
Cutting to hips	m	0.20
Ridge and hip tiles	m	0.15

Although items such as cutting tiles are deemed to be included, an allowance will have to be made in the pricing to cover this.

In order to form boundary work, tile manufacturers produce purpose-made tiles. Many of these fittings are dry fixed with clips, while others have to be bedded in cement mortar, for example:

- 165 × 215 mm eaves tile;
- 248 × 265 mm tile and half verge tile.

Example

260 × 305 mm long half round ridge bedded in cement mortar (1:3). Cost per m.

Data

260 × 305 mm half round ridge tile £3.50 each
Assume 10 tiles

Materials

10 ridge tiles @ £3.50 each	35.00
Add waste 10%	3.50
Bedding 0.03 m^3 per m = 0.09 m^3 mortar @ £116.92 m^3	10.52
Add waste 5%	0.53
	49.55

Labour

0.15 hour per m = 0.45 hours @ 23.16		
	15.75	
	38.91	17.51
		67.06
Add profit and overheads 15%		10.06
Cost per 10 tiles (3 m)		77.12
÷ 3 – cost per m		£25.71

Example

Boundary work at eaves including 165 × 215 mm eaves tile.

With double-lap plain tiling, it is necessary to use a special eaves tile in order to provide the lap and gauge for the rest of the roof; typically an eaves tile is 165 × 215 mm, that's to say 50 mm shorter than a normal tile. When an extra eaves course is used it is also necessary to have an extra batten to provide for fixing the eaves tile.

Data

165 × 215 mm eaves tile £1.20 each
Batten (see calculation above) @ £3.20 per 10 m
Assume 10 m

£

Materials

$\dfrac{10,000}{165} = 61 \times £1.20$ 73.20

Add waste 10%	7.32
10 m batten	3.20
Add waste 5%	0.15
Nails for battens and eaves tiles, say	<u>2.00</u>
	85.87

Labour

10 m @ 0.05 = 0.50 hours @ £38.91	<u>19.46</u>
	105.33
Add profit and overheads 15%	<u>15.80</u>
Cost per 10 m	<u>121.13</u>
÷ 10 – cost per m	<u>£12.11</u>

Example

Boundary work at verge including 248 × 265 mm tile and half.

In a similar fashion to special eaves tiles, verge tiles are special tiles, being 123 mm wider than normal roofing tiles in order provide the double-lap coverage.

Note: no additional batten is required for this item.

165 × 248 mm verge tile £1.50 each
Assume 10 m

£

Materials

$\dfrac{10,000}{248}$ = say 40 verge tiles @ £1.50	60.00
Add waste 10%	6.00
Nails, say	<u>1.00</u>
	67.00

Labour

10 m @ 0.05 = 0.50 hours @ £38.91	<u>19.46</u>
	86.46
Add profit and overheads 15%	<u>12.97</u>
Cost per 10 m	<u>99.43</u>
÷ 10 – cost per m	<u>£9.94</u>

Single-lap tiles

Interlocking tiles are single-lap tiles that are fixed with nails or clips. They are generally made from concrete, with an applied finish. They are heavier than slate or clay tiles and therefore should be used with caution when used as a replacement to lighter, more traditional coverings on an existing roof. Their extra weight is a disadvantage from the site handling point of view; however, they have greater coverage than slates or tiles and can be fixed more quickly, there being 9.8 interlocking tiles per square metre compared with 60 for double-lap clay tiles, and these factors combine to provide a cheaper alternative that is popular with speculative house builders.

Single-lap tiles are laid to 345 mm gauge as manufacturer's recommendations; therefore, battens required are 2.90 m per m².

Example

Roof coverings, 420 × 330 mm Marley Ludlow Major Antique Brown concrete interlocking tiles with 75 mm head lap 40° pitch with 38 × 25 mm treated sawn softwood battens at 345 mm centres fixed with clips including one layer of felt to BS Type 1F with 100 mm minimum laps fixed with galvanised clout nails. Cost per m².

Materials delivered to site

Malvern concrete interlocking tiles	£696.00 per 1,000
Underfelt	£13.00 per 15 m × 1 m roll
Nails for tiles	£2.00 per kg
Battens	£3.20 per 10 m
Nails for battens	£1.00 per kg

Assume 10 m²	£	£
Tiles – from Marley web site – 9.8 per m²:		
98 Malvern tiles @ £696.00 per 1,000	68.21	
Add waste 10%	6.82	
Battens:		
29 m batten @ £3.20 / 10m	9.28	
Add waste 10%	0.93	
Nails for battens:		
0.40 kg @ £1.00 per kg	0.40	
Add waste 5%	0.02	
Nails for tiles – nailed every fourth course:		
$\dfrac{98}{4} = 25$ nails say 30 to include waste, say	0.20	85.86

Underfelt:

15×1 m roll = 15 m^2

$$\text{Cost per 10 m}^2 = \frac{10.00}{15.00} \times £\,13.00 \qquad\qquad 8.67$$

Add laps 10%	0.87	
Add waste 5%	<u>0.44</u>	95.84

Labour

Tiler and labourer can lay 3.5 m^2 of interlocking tiles per hour

Tiles:

2.86 hours tiler and labourer	£23.16		
	£15.75	38.91	111.28

Tile battens:

0.30 hours tiler	23.16	6.95
0.15 hours labourer	15.75	<u>2.36</u>
		£216.43
Add profit and overheads 15%		<u>£32.47</u>
		£248.90
$\div$ 10 – cost per m^2		<u>£24.89</u>

SECTION 19: WATER PROOFING

Mastic asphalt roofing

Asphalt is always applied on sheathing felt to isolate it from the structure and this has to be included in the calculation. The finished surface must be protected against deterioration due to exposure to solar radiation; this is usually in the form of solar-reflecting paint in the case of asphalt.

Labour constants – craftsman and labourer

19 mm two coat asphalt roofing   0.15 per m^2
Skirtings, fascias and aprons 150–225 mm girth (including internal angle fillet and labours)   0.60 per m

Example

19 mm thick two-coat asphalt roof covering exceeding 500 mm wide to 5° slope with and including one layer of sheathing felt and 50 mm thick rigid glass

fibre insulation to concrete, finished with two coats of solar reflecting paint. Cost per m².

Asphalt is manufactured in 25 kg blocks and then heated to melting point and applied on site. The following is the approximate covering capacity of 1,000 kg of asphalt:

- first 12 mm thickness: 35 m²;
- additional 3 mm thickness: 150 m².

For example, the following is the weight of asphalt required per square metre 19 mm thick:

$$\text{first } 12\text{mm} = \frac{1000\,\text{kg}}{35\,\text{square metres}} = \qquad 28.57\,\text{kg}$$

$$\text{next } 6\,\text{mm} = \frac{1000\,\text{kg}}{150\,\text{square metres}} \times \frac{7\,\text{mm}}{3\,\text{mm}} = \underline{15.56\,\text{kg}}$$

$$\underline{44.13\,\text{kg}}$$

Data

Assume 10 m²

£

Materials

	£
10 m² × 44.13 kg = 441.3 kg mastic asphalt in pots @ £300.00 per tonne	132.39
10 m² sheathing felt @ £1.50/m²	15.00
10 m² 50 thick rigid glass fibre insulation @ £2.50/m²	25.00
10 m² solar reflecting paint @ £1.00/m²	10.00
	£182.39
Add waste 5%	£9.12
	£191.51

Labour

Asphalt:

Spreader @ £23.16
Labourer @ £15.75
Gang rate £38.91 per hour

0.3 hour gang rate = 3 hours @ 38.91	116.73

Insulation:

0.2 hour/m^2 carpenter = 2 hours @ £23.16	46.32

Painting:

0.1 hour/m^2 spreader – 1 hour @ £23.16	<u>23.16</u>
	377.72
Add profit and overheads 15%	<u>56.66</u>
Cost per 10 m^2	<u>434.38</u>
÷ 10 – cost per m^2	<u>£43.44</u>

Sundry labour constants for asphalt work – based on gang rates as above

Item	Gang hours per m^2
One coat horizontal over 500 mm wide	0.30
Two coat vertical over 500 mm wide	1.00

	Gang hours per m
Skirtings, fascias and aprons 150 mm girth	0.30
Two-coat lining to gutter 450 mm girth	1.00
Two-coat lining to 300 × 300 × 200 mm deep outlet	1.10

SECTION 20: PROPRIETARY LININGS AND PARTITIONS

Proprietary linings and partitions are used in situations where existing walls need to be lined on the inside face or where metal partitions are used in preference to timber studs due to their greater flexibility, stability or speed of erection. Systems comprise metal studs and channels that are fixed to the structure and then faced either on one or both sides with plasterboard. The systems can also be insulated and fitted with vapour barriers as illustrated in Figures 4.13 and 4.14.

Example

100 mm finished thickness Knauf stud and track partitioning system comprising 70 mm wide stud and track plugged and screwed to masonry and concrete to form walls over 2 m not exceeding 3 m high, finished on both sides with 9.5 mm tapered plasterboard with taped joints fixed with drywall screws. Cost per m^2.

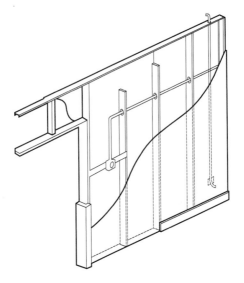

Figure 4.13 Metal proprietary partitioning

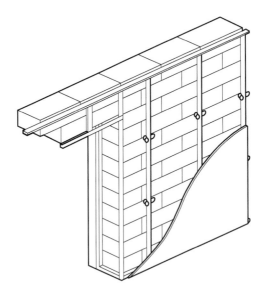

Figure 4.14 Proprietary systems used as wall lining

Example

40 mm thick overall Gyplyner wall lining system comprising 34 × 50 mm Gyplyner metal track an lining channels plugged and screwed to masonry Gypframe brackets and finished with 12.5 mm Gyproc tapered edge wallboard one side fixed with drywall screws and taped joints. Cost per m².

As the name of the section suggests, the systems illustrated in Figures 4.13 and 4.14 are proprietary systems and as such will be supplied and fixed by a sub-contractor. The systems, once installed, are finished with plasterboard on one or both sides with drywall screws at rates in line with the examples in Section 28, later in this chapter.

SECTION 21: CLADDING AND COVERING

This section of NRM2 includes a number of trades that are usually carried out by specialist sub-contractors and may be the subject of work packages rather than a bill of quantities. However, although prices may be supplied by specialists, it is useful for the estimator to have a working knowledge of the systems being quoted for.

Rainscreen

A rainscreen façade is a cladding applied either during primary construction or as an over cladding to an existing structure. Rainscreen cladding consists of an outer weather-resistant decorative skin fixed to an underlying structure by means of a supporting grid, which maintains a ventilated and drained cavity between the façade and the structure. A range of metal and metal composite materials (MCMs) are used to manufacture rainscreen cladding systems. MCMs consist of two thin skins of aluminium or other metals such as copper, zinc and stainless steel continuously bonded under tension to either side of a thermoplastic or mineral core.

Rainscreen façades are not normally sealed and a ventilation cavity of at least 25 mm is allowed immediately behind the cladding panel. Insulation can be positioned within the cavity and openings at the top and bottom of clad areas allow for evaporation of moisture vapour and ventilation/drainage. A ventilated rainscreen incorporating insulation will allow the building fabric to breathe without the risk of interstitial condensation or structural decay. External wall insulation used in this way is superior in performance as it eliminates the condensation risks associated with internal or cavity-wall insulation. This is particularly important for refurbishment schemes. In new construction, the use of back ventilated rainscreen cladding provides the designer with the opportunity to use economical single-skin load-bearing blockwork for infill walls. The need for complicated damp-proof membrane detailing is eliminated and there will be less risk of cold bridging. Where lower performance is required, for example in low-rise structures, then a similar cassette panel system, although not fully pressure equalised, performs well.

Profiled sheet cladding/roofing

Profiled sheet cladding/roofing is a very popular material for a variety of buildings from domestic to industrial (Figure 4.15). The sheets must be lapped at ends and sides and there is a number of accessories available for use at abutments and boundaries. Sheets are available with integral insulation.

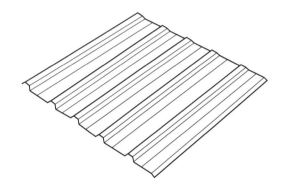

Figure 4.15 Profiled sheeting

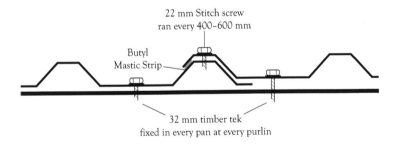

Figure 4.16 Fixing profile sheeting

If fixing into timber, 32 mm timber teks should be used and are fixed through the pan (Figure 4.16). Tek fixings are self-cutting and sealing; however, it is advisable to drill a small pilot hole first. 22 mm Stitch screws can also be used to pinch the overlaps together; this is recommended on roofs with a shallow pitch. It is also recommended to use a Butyl sealant tape on all overlaps for shallow-pitch roofs. Teks is the generic name for the fittings that should be used to attach the sheets. When using teks they must always be fitted at a 90-degree angle to the sheet and secured tight enough, until there is a slight bulge; however, too much tightening may cause damage to the sheets and cause them to leak.

Example

0.5 mm thick polyester-coated galvanised steel profile roofing to 10° pitch fixed to timber with screw, washer and cap. Cost per m^2.

Data

Profiled steel roofing sheets come in a variety of sizes but each sheet has an effective width of 1 m. At the sides there should a minimum of 300 mm lap.

3 m long × 1.30 m (effective width 1.00 m) steel profile roofing £32.00 per sheet
Pack of 100 tek fixings £20.00
Assume 27 m^2

Materials

Note that the quantities in the bill of quantities/work package will be measured net and this should be allowed for when pricing.

   3 m long × 1 m effective width steel profile roofing – an allowance of 300 mm (150 mm for each end) for laps has to be made, therefore the effective roof coverage of a 3 × 1m sheet will be:

   2.70 × 1.00 = 2.70 m^2

	£
27 profile sheet roofing @ £32.00 per sheet	864.00
Tek fixings 20 per sheet × 27 sheets – 540 @ £20/100	108.00
Butyl sealant, say	15.00
	987.00
Add waste 10%	98.70
	1,085.87

Labour

0.6 hours/m^2 roofer and labourer

27 m^2 × 0.60 = 16.2 hours @ Roofer	@ £23.16	
Labourer @ £15.75		
Gang rate £38.91 per hour		630.34
		1,716.21
Add profit and overheads 15%		257.43
Cost per 27 m^2		1,973.64
÷ 27 – cost per m^2		£73.10

SECTION 22: GENERAL JOINERY

- Unframed isolated trims, skirtings, or sundry joinery items.
- In-fill panels and sheets.
- Sealant joints.
- General ironmongery not associated with windows and doors.

Sundry labour constants

Item	Hours (joiner/m)
Gutter boarding	1.00
19 mm fascia	1.00
Barge board	0.40
19 × 50 mm architrave	0.10
19 × 75 mm skirting	0.15
Window board	0.30
Extra for plugging to brickwork	0.20

Allow 0.05 kg nails per m.

Example

14.5 × 44 mm painted MDF chamfered architrave. Cost per m.

Data

14.5 × 44 mm painted MDF chamfered architrave delivered to site £4.00/2.50 m
Fixings delivered to site £2.00/kg
Assume 10 m

Materials

10 m 14.5 × 44 mm MDF architrave @ £4.00/2.5 m	£16.00
Nails 0.5 kg nails @ £2.00/kg	£1.00
	£17.00
Add waste 7%	£1.19
	£18.19

Labour

1 hour carpenter/joiner @ £23.16	£23.16
	£41.35
Add profit and overheads 15%	£6.20

| Cost per 10 m | £47.55 |
| ÷ 10 – cost per m | £4.76 |

Example

244 × 25 mm painted MDF window board with rounded front edge. Cost per m.

Data

244 × 25 mm painted MDF window board delivered to site £38.00/3.66 m length
Fixings £2.00/kg
Assume 10 m

Materials

$$\frac{10.00}{3.66} = 2.73$$

2.73 m MDF window board @ £38.00/3.66 m	£28.34
0.5 kg fixings @ £2.00 per kg	£1.00
	£29.34
Add waste 7.5%	£2.20
	£31.54

Labour

3 hours joiner @ £23.16	£69.48
	£101.02
Add profit and overheads 15%	£15.15
Cost per 10 m	£126.17
÷ 10 – cost per m	£12.62

General ironmongery not associated with windows and doors

Example

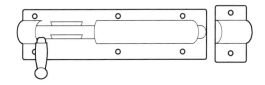

Figure 4.17 Barrel bolt

300 mm black japanned barrel bolt and fixing to softwood (Figure 4.17). Cost per no.

Materials

1 no. black japanned barrel bolt @ £4.25 (screws included)	4.25

Labour

0.60 joiner @ £23.16	<u>13.90</u>
	18.15
Add profit and overheads 15%	<u>2.72</u>
Cost per no.	<u>£20.87</u>

Sundry labour constants

Description	Joiner (hours per unit)
Small numeral	0.10
Cupboard catch	0.25
Rubber door stop to concrete	0.20
Hat and coat hook	0.15
100 × 150 mm shelf bracket	0.25
75 mm hasp and staple (Figure 4.18)	0.25
Suffolk latch	1.00

The above constants are for fixing to softwood; for fixing to hardwood, add 20%.

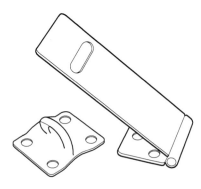

Figure 4.18 Hasp and staple

SECTION 23: WINDOWS, SCREENS AND LIGHTS

In this section, note that bedding and pointing frames and glazing are deemed to be included.

Example

630 × 1050 mm Sovereign Stormsure (LEW112CAS) softwood factory-glazed window. Cost per no.

Data

625 × 1195 mm Stormsure (LEW112CAS) softwood window, factory glazed including ironmongery £340.00
Mastic pointing £5.38 per tube

Materials	£
625 × 1195 mm Stormsure softwood window	340.00
Pointing 625 + 1,195 = 1,820 × 2 = 3,640 m × 0.1 = 0.36	
0.36 tube of mastic @ £5.38 per tube	<u>1.94</u>
	341.94

Labour

Fixing wood 625 × 1195 mm window 0.7 hour joiner @ £23.16	16.21
Bedding and pointing 0.18 hour per m × 3.64 = 0.66 @ £23.16	<u>15.29</u>
	373.44
Add profit and overheads 15%	<u>56.02</u>
Cost per no.	<u>£429.46</u>

Sundry constants

Description	Joiner (hours per unit)
Unloading, stacking, hoisting and fixing wood windows; 630 × 750 mm	0.40
Ditto 630 × 1050 mm	0.70
Ditto 1200 × 900 mm	0.90
Ditto 1200 × 1200 mm	1.20
Ditto 2339 × 1200 mm	2.30
Pointing one side in mastic	0.18 plus 0.10 tube of mastic per linear metre

Metal windows

Example

1830 × 1560 mm Crittal Homelight polyester powder-coated aluminium window with satin chrome furniture, side hung, double glazed, fixed with screws to hardwood sub-frame (measured separately). Cost per no.

Data

1850 × 1560 mm aluminium window £1,600.00
Note that windows are delivered to site covered in tape to protect from damage during transit and fixing in place. Whereas removal and disposal of this protection may not be an issue with individual units, it can be on large projects where many hundreds of units are involved.

Materials	£
1 no. 1850 × 1560 mm aluminium window	1,600.00
Mastic and fixings, say	<u>10.00</u>
	1,610.00

Labour	
Unloading, storing, removing and disposing of protective tape, 2 hours joiner @ £23.16	<u>46.32</u>
	1,656.32
Add profit and overheads 15%	<u>248.45</u>
Cost per unit	<u>£1,904.77</u>

SECTION 24: DOORS, SHUTTERS AND HATCHES

Doors, shutters and hatches and associated ironmongery.

Sundry labour constants

Doors – allowances include for fixing and hanging on standard hinges:

Item	Hours (joiner)
40 mm standard door	1.00
50 mm solid-core door	1.40
44 mm panelled door	1.25

Example

44 × 762 × 1981 mm oak-veneered flush external door £571.00

Materials

44 × 762 × 1981 mm external door	£571.00

Labour

1.40 hours @ £23.16	£32.42
	£603.42
Add profit and overheads 15%	£90.51
Cost per door – supplied and fixed	£693.93

Door linings can be measured either in linear metres or as a set. If measured in linear metres then the door stop will be measured separately.

Example

32 × 132 mm softwood door lining, plugged and screwed to masonry. Cost per m. Assume 10 linear metres.

Data

32 × 132 mm softwood lining £4.35 per m

Materials

10 m 32 × 132 mm softwood door lining @ £4.35/m	43.50
Add screws and plugs, say	0.50
	44.00
Add waste 7.5%	3.30
	47.30

Labour

0.10 hours/m + 0.20 hours/m plugging = 0.30 hours/m	
10 × 0.30 = 3 hours @ £23.16	69.48
	116.78
Add profit and overheads 15%	17.52
Cost per 10 m	134.30
÷10 – cost per m	£13.43

Example

32 × 132 mm softwood door-lining set, plugged and screwed to masonry. Cost per no.

Data

32 × 132 mm softwood-lining set £33.18

Materials

1 no. 32 × 132 mm lining set	33.18
Add plugs and screws, say	2.50
	35.68
Add waste 7.5%	2.68
	38.36

Labour

1.5 hours @ £23.16 per hour	34.74
	73.10
Add profit and overheads 15%	10.97
Cost per lining set	£84.07

SECTION 25: STAIRS, WALKWAYS AND BALUSTRADES

Example

Straight-flight pine staircase to BS 585: Part 1, compliant: Part K and M; 855 mm wide, total going 2600 mm with 237 × 32 mm strings, 241 × 22 mm treads, 82 × 82 × 900 mm newel posts top and bottom, newel caps, 63 × 44mm handrail, 32 × 32 mm balusters at 95 mm centres; one string plugged to wall, other string housed both ends to newels, left clean for clear finish.

Data

Pre-fabricated staircase £436.00

Materials

1 no. pre-fabricated pine staircase	436.00
Add sundry fixings – nails and bolts, say	20.00
	456.00

Labour

5 hours joiner to fit and fix straight flight @ £23.16	<u>115.80</u>
	571.80
Add profit and overheads 15%	<u>85.77</u>
Cost per no.	<u>£657.57</u>

SECTION 26: METALWORK

Example

1350 mm long Catnic CG110/125 standard duty lintel bedded on blockwork in gauged mortar (1:1:6). Cost per no.

Data

CG110/125 Catnic cavity-wall lintel £110.60 each

Materials

1 no. Catnic cavity-wall lintel	110.60
Mortar for bedding in place, say	<u>3.00</u>
	113.60

Labour

0.3 hours bricklayer @ £23.16	<u>6.95</u>
	120.55
Add profit and overheads 15%	<u>18.08</u>
Cost per no.	<u>£138.63</u>

SECTION 27: GLAZING

Until comparatively recently, glass was either described as sheet or float, reflecting the way in which it was manufactured but today glass has become much more sophisticated and is manufactured in a number of different forms in order to meet the need to provide the following:

- Solar control: glass provides a more comfortable environment, and can help to reduce the capital outlay, running costs and carbon emissions of a building throughout the year, by reducing or eliminating the requirement for air conditioning.

- Thermal control: with increasing environmental awareness, more emphasis is now being placed on ways to save energy in any building, domestic or commercial – to the extent that, in recent years, new regulations have been introduced specifying minimum requirements for energy efficiency (Part L of the Building Regulations) – and glass can play an important role in saving energy.
- Fire protection: fire-resistant glass offers varying levels of protection, which is measured in terms of integrity (the time the glass remains intact in a fire) and insulation (the amount the glass will limit the temperature rise on the non-fire side).
- Noise protection: laminated glass can reduce noise from a variety of external sources including traffic on the road, rail and in the air, in order to decrease stress and noise-related illnesses.
- Security: toughened safety glass is up to five times stronger than ordinary annealed float glass of the same thickness. It is mandatory to use safety glass in certain situations.
- Self-cleaning: this glass has a specially developed coating on the outside that once exposed to daylight reacts in two ways: first, it breaks down any 'organic' dirt deposits such as the organic content of bird droppings and tree sap; and second, rain water 'sheets' down the glass to wash the loosened dirt away.
- Decoration and texture: to add privacy or meet any other aesthetic or practical requirements.

The requirements of NRM2

Glass in doors and window can be measured and billed in Section 23 or 24 or, if glazing is not associated with these items, it can be measured and included in this section. NRM2 requires that glass is measured and billed as a numbered item.

Sundry constants

Although NRM2 requires that glazing is numbered, the following constants reflect rates per m²:

Item	Labour/m² (to be converted to no. for NRM2)
3 mm clear float glass and glazing to wood	1.25 hours
5 mm Ditto	1.25 hours
6 mm wired glass	1.70 hours
Mirrors	1.00 hour

Waste 10% or 5% on off-site cut-to-size glass.

Amount of glazing compound:

Item	Kg/m²
Back and front putty including pointing	2.00
Back putty only and bedding glazing beads	0.70

Example

4 mm clear float glass in pane size 900 × 900 mm and glazing to wood with glazing compound. Cost per no.

Data

900 × 900 mm 4 mm clear float glass cut to size and delivered to site £31.59
Glazing compound £5.99/2 kg

	£

Materials

900 × 900 mm pane, clear sheet glass, cut to size	31.59
Putty 0.9 × 0.9 = 0.81/m² × 2.00 = 1.62 kg putty @ £5.99 per 2 kg	<u>4.85</u>
	36.44
Add waste 5%	<u>1.82</u>
	38.26

Labour

0.81 × 1.25 = 1.01 hours glazer @ £23.16	<u>23.39</u>
	61.65
Add profit and overheads 15%	<u>9.25</u>
Cost per 900 × 900 mm pane	<u>£70.90</u>

SECTION 28: FLOOR, WALL, CEILING AND ROOF FINISHINGS

In-situ, tiled, block, mosaic, sheet, applied liquid or planted finishes.

In-situ finishes

Most modern in-situ plasterwork is carried out in lightweight gypsum plaster. It is known by its trade names Carlite or Thistle, although Carlite has now been merged into the Thistle brand. It is delivered to site in 25 kg bags on pallets for ease of unloading. These retarded hemihydrate plasters are premixed and contain lightweight aggregate and gypsum plaster, and require only the addition of water to make them ready for use. Thistle plasters come in a variety of types depending on where they are to be used and the background to which the plaster is applied. The finish grades are used neat, while the undercoat grades are usually mixed with sand. NRM2 states work over 600 mm wide is measured in square metres while work under 600 mm wide is measured in linear metres.

It is very important that the correct type of plaster is used; the five types (four undercoats and one topcoat) available are as follows:

Undercoats

- **Browning plaster** is an undercoat plaster for moderate-suction solid backgrounds that have a good mechanical key such as brickwork or blockwork. A slow-setting variety is available that give greater time for application.
- **Bonding plaster** is an undercoat plaster for low-suction backgrounds, for example concrete, plasterboard or surfaces sealed with PVA (a universal water-based adhesive).
- **Toughcoat** is an undercoat plaster for solid backgrounds of high suction with an adequate mechanical key.
- **Hardwall** is an undercoat plaster that provides a much harder and more durable finish and is also quick drying.

Finishing coats

- **Finishing plaster** is an idea choice over sand and cement bases and can be used on still-damp backgrounds.
- **Board finishing plaster** is a one-coat plaster for skim coats to plasterboard.
- **Multi finish** is used where both undercoat and skim coat are needed on one job. It is suitable for all suction backgrounds and ideal for amateur plasterers.

In addition the following is available:

- **Universal one coat** is a one-coat plaster for a variety of backgrounds, suitable for application by hand or machine.

The White Book published by British Gypsum is a good source of reference for all types of plaster finishes. It can be viewed at: www.british-gypsum.com/literature/white-book.

What is used in practice?

- For plasterboard ceilings: board finish plaster.
- For blockwork walls: an 11 mm thick undercoat of browning plaster followed by 2 mm skim coat of finishing plaster. The undercoat is lightly scratched to form a key.

- Metal lathing: two undercoats are often required followed by a finishing coat.
- Galvanised or stainless steel angle beads are used to form external angles in in-situ plaster.
- In situations where damp walls are plastered following the installation of an injection damp-proof course, most gypsum plasters are not suitable as they absorb water and fail.

Approximate coverage per tonne	m^2
11 mm Browning plaster	135–155
11 mm toughcoat plaster	135–150
11 mm hardwall plaster	115–130
2 mm board finishing plaster	410–430
2 mm finishing plaster	410–430
13 mm universal one coat	85–95

Labour constants

Item	Plasterer and labourer / 10m^2
Base coat	2 hours
Finishing coat	2.5 hours

For ceilings, increase by 25%.

Sand/cement screeds

The function of a floor screed is to provide a smooth and even surface for finishes such as tiling and it is usually a mixture of cement and course sand, typically in the ratio of 1:3. It is usually a dry mix with the minimum of water added and has a thickness of between 38–50 mm and is laid on top of the structural floor slab.

Board finishes

Plasterboard is available in a variety of thicknesses, the most common being 9.5 mm and 12 mm, and is fixed to either metal of timber studs with screws. Standard sheets are 900 × 1800 mm and 1200 × 2400 mm. It is also possible to fix plasterboard to blockwork with plaster dabs. Boards can either be pre-finished and require no further work or finished with a skim coat of plaster. Boards with tapered edges and foil backing are also available.

Example

13 mm thick two-coat plaster comprising 11 mm thick Thistle bonding coat and 2
mm thick Thistle multi-finish plaster on block walls over 600 mm wide internally
with steel-trowel finish. Cost per m².
Assume 10 m².

Data

Bonding plaster £431.60 per tonne delivered to site on pallet and unloaded
Multi-finish plaster £302.80 per tonne delivered to site on pallet and unloaded

Materials

Approximate coverage of bonding plaster 11 mm thick – 100m²/tonne
Approximate coverage of multi-finish plaster 2 mm thick – 420m²/tonne
Bonding plaster:

$$\frac{10.00}{100.00} \times £431.60 =$$ £43.16

Finishing plaster:

$$\frac{10.00}{420.00} \times £302.80 =$$ £7.21

                                                                                    £50.37
Add waste 10%   £5.04
                                                                                    £55.41

Labour

Plasterer and labourer:

Base coat	2.00 hours	
Finishing coat	2.50 hours	
	4.50 hours @ £23.16	
	£15.75	
	£38.91	£175.10

                                                                                    £230.51
Add profit and overheads 15%   £34.58
Cost per 10 m²   £265.09
÷ 10 – cost per m²   £26.51

Floor and wall tiling

Floor tiling

When laying inflexible floor finishes such as ceramic or quarry tiles, the structural surface is covered with a screeded bed. It is important that the surface of the screed on which the tiles are to be laid is true and even, therefore a steel-trowel finish is required, measureable in accordance with NRM2. For more flexible finishes and carpet, the same degree of accuracy is not normally required. NRM2 states that the screed and tiles are measured separately and that all rough and fair square, raking and curved cutting, waste and raking out joint to form a key are deemed to be included and not measured separately.

Example

Screeded bed.

38 mm thick cement and sand (1:3) screed exceeding 600 mm wide level and to falls only not exceeding 15˚ from horizontal to concrete with steel-trowel finish. Cost per m².

Materials

	£	£
1 m³ cement = 1,400 kg cement @ £190.00 per tonne	266.00	
3 m³ sand = 4,800 kg @ £15.00 per tonne	72.00	
	338.00	
Add shrinkage 25%	84.50	
	422.50	
Add waste 5%	21.13	
Cost per 4 m³		
	£433.63	
÷ 4 – cost per m³	£108.41	108.41

Mixing

Assume 100 litre mixer @ £24.00 per hour		
Output 4 m³ per hour – cost per m³		6.00
Cost per m³		114.41

Cost per m² 12 mm thick $\dfrac{£114.41}{1} \times \dfrac{38}{1,000} =$ 4.35

Labour

Plasterer @ £23.16
Labourer @ £15.75
Gang rate £38.91 per hour
0.5 hour gang rate @ £38.91   <u>19.46</u>
                                                        23.81
Add profit and overheads 15%   <u>3.57</u>
Cost per m²   <u>£27.38</u>

Example

331 × 331 × 8 mm BCT white satin floor tiles exceeding 600 mm wide laid on
screeded bed to chessboard pattern bedded and pointed with BAL adhesive.
Assume 10 m².

Data

BCT floor tiles £27.38/m²
BAL tile adhesive and grout £66.00 per 10 litres
5 m² per 10 litres 1 mm thick adhesive including grout

Materials	**£**
10 m² Padova floor tiles @ £27.38/m²	273.80
Add waste 10%	<u>27.38</u>
	301.18
20 litres BAL tile adhesive and grout @ £66 per 10 litres	132.00
Add waste 5%	<u>6.60</u>
	439.78

Labour

0.5 hours/m² = 5 hours @ £23.16	<u>115.80</u>
	555.58
Add profit and overheads 15%	<u>83.34</u>
Cost per 10 m²	<u>£638.92</u>
÷ 10 – cost per m²	<u>£63.89</u>

Wall tiling

Most wall tiles are 100 × 100 mm with spacers to ensure ease of fixing. Tiles require
adhesive for fixing and grout for the joints. Thickness can vary and floor tiles can be
found in a wide variety of sizes.

Example

223 × 223 × 10 mm thick white rustic wall tiles exceeding 600 mm wide fixed to plaster with tile adhesive including grouting joints as work proceeds. Cost per m².

Data

White rustic wall tiles £21.81 per m²
Unibond wall tile adhesive and grout £15.60 per 10 litres
5 m² per 10 litres 1 mm thick adhesive including grout
Assume 10 m²

Materials

	£
10 m² white rustic wall tiles @ £21.81/m²	218.10
Add waste 10%	21.81
Adhesive and grout:	
20 litres @ £15.60 per 10 litres	31.20
Add waste 5%	_7.80_
	278.91

Labour

0.6 hour per m² = 6 hours @ £23.16	_138.96_
	417.87
Add profit and overheads 15%	_62.68_
Cost per 10 m²	£480.55
÷ 10 – cost per m²	£48.06

Example

152 × 152 × 5.5 mm thick Johnson opal white glazed tiles to plasterboard walls exceeding 600 mm wide bedded in BAL white waterproof adhesive and grouting joints as work proceeds. Cost per m².

Data

White rustic wall tiles £27.49 per pack (43 tiles)
Unibond wall tile adhesive and grout £15.60 per 5 litres
Coverage:
5 m² of tiles per 10 litres 1 mm thick adhesive including grout
Assume 1 m²

Materials

	£
Tiles:	
Coverage 43 tiles per m²	
43 tiles @ £27.49 per pack	27.49
Add waste 10%	2.75
Adhesive and grout:	
2 litres @ £15.60 per 5 litres	6.24
Add waste 5%	0.31
	36.79

Labour

	£
0.6 hour per m² @ £23.16	13.90
	50.69
Add profit and overheads 15%	7.60
Cost per m²	58.29

Sundry labour constants

Description	Unit	Labour (hours)
Metal angle bead	m	0.10
Make good new plaster to old	m	0.15
Vinyl sheet laid to stair treads not exceeding 600 mm	m	0.25

SECTION 29: DECORATION

Decoration or, more particularly, painting is one of the building trades that has changed radically over the last 30–40 years, with the introduction of painting systems rather than oil-based paints. In its widest sense, decoration includes painting on a variety of surfaces and paper hanging.

The traditional approach to painting wood is 'KPS 3 oils'; that is, knot, prime and stop, and apply two undercoats and one gloss coat to external surfaces and one undercoat and one gloss coat to internal surfaces. Knotting solution is a shellac/methylated spirit-based sealer that is applied to knots in the wood, thereby preventing the sap in the knot from seeping out and bubbling through the paint finish, blistering the surface. The surface is primed by applying a coat of primer to the bare wood. Traditionally, stopping refers to a linseed-oil putty being applied to all small holes and cracks in the wood. There is a popular belief that paint fills small holes and cracks, but that's not the case.

All these products are still available today but paints have moved away from being oil based to solvent based or water based, and wood filler is used in preference to putty as it does not dry out and crack so easily.

In practice, it is now usual to apply a coat of primer or combined primer/undercoat followed by only one gloss coat whether internal or external, except in very exposed or extreme conditions.

Increasingly in some locations – public staircases etc. – paint is applied by spraying. Portaflek and Multiflek finishes were very popular in the 1970s, but now they have been superseded by compliant water-based versions: Aquaflek and Aquatone. The advantage of spray application is that it is considerably quicker than using a brush or roller; the disadvantages are that a considerable amount of time is required for masking, protection and cleaning.

When pricing painting and decorating, it is important to study the specification or preamble to the section as this will contain information relating to the preparation work that is expected. More than most trades, this can be extensive and includes washing down, burning off existing paint and stripping existing paper, etc.

Two peculiarities of estimating painting and decorating are: brush money, an allowance given to cover the cost of new brushes, rollers, etc.; and an allowance to cover cleaning overalls etc. Both of these allowances are covered in the National Working Rule Agreement mentioned previously in this chapter.

Sundry labour constants

Item	Skilled hours per 100 m²
Knotting and stopping	4.50
Priming	11.00
Undercoat	14.00
Gloss coat	18.00
Emulsion paint	8.00
Preparation of surfaces (rubbing down between coats, etc.)	4.00
Creosoting timber	18.00
Wax polishing	50.00

Add to the above for:

Description	% increase (labour)	% increase (materials)
Not exceeding 300 mm girth	25	10
Work over 3.5 m above floor level	10	0

Example

Prepare and two coats of emulsion on general surfaces of plaster internal. Cost per m².

Data

Dulux trade emulsion paint £39.17 per 5.0 litres
Assume 100 m²

Materials

	£
Coverage:	
First coat 15 m² per litre 100 ÷ 15 = 6.67	
Second coat 8 m² per litre 100 ÷ 8 = <u>12.50</u>	
<u>19.17, say 19 litres</u>	
19 litres of emulsion @ £39.17 per 5.0 litres	148.85

Labour

Preparation	5 hours	
Emulsion (2 coats)	16 hours = 21 hours @ £23.16	<u>486.36</u>
		635.21
Add profit and overheads 15%		<u>95.28</u>
Cost per 100 m²		<u>730.49</u>
÷ 100 – cost per m²		<u>£7.31</u>

Decorative papers or fabrics

Wallpaper

Wallpaper is specified in 'pieces', more commonly called a roll. The standard UK roll or piece is 0.53 m wide × 10.05 m long. Some European papers are a little narrower and many American rolls are roughly twice as wide. The amount of paper required will depend on the pattern, the size of the room and the number of openings, etc. Once again, the surface to be papered may need to be prepared. For example, new plasterboard will require at least two coats of drywall primer to adjust the suction, whereas old plastered walls can become very dry and powdery, often referred to as 'blown' and need a coat of plaster sealer. Paper hanging to ceiling takes approximately 50% longer than walls! Newly plastered walls will required a coat of size, usually dilute paper adhesive, to adjust the suction. It is common to have wallpaper specified in the bills of quantities as a prime cost or provisional sum and the labour for hanging measured as a hanging/fix only item.

Waste on wallpaper will vary according to the pattern, the number of window and door openings in the room and the height of the room. The following is the average:

Item	% waste
Lining paper to walls	7.50
Patterned paper to walls	15.00
Lining paper to ceiling	5.00
Plain paper to ceilings	7.50
Borders and strips	7.50

Sundry labour constants

Item	Skilled hours per 10 m^2
Strip off old paper to walls	1.00
Applying size and preparation	0.50
Lining paper to walls	2.00
Patterned paper to walls	2.50
Lining paper to ceilings	2.50
Patterned paper to ceilings	3.00

A roll of UK wallpaper is 0.53 m wide and 10.05 m long, which equals 5.33 m^2.

Example

Strip existing wall coverings, prepare, and hang patterned wallpaper (PC £20.00 per piece delivered to site) to walls. Cost per m^2.

Materials

Wallpaper per roll	£20.00	
Add waste 15%	£3.00	
	£23.00 ÷ 5.33 = £4.32/m^2	
Assume 10 m^2		
10 m^2 patterned wallpaper @ £4.32/m^2		43.20
Paste, size and caulk		2.00
		45.20

Labour

Strip old paper	1.00
Prepare walls/size	0.50

Hang paper	2.50	
	4.00 hours painter @ £23.16	92.64
		137.84
Add profit and overheads 15%		20.68
Cost per 10 m²		158.52
÷ 10 – cost per m²		£15.85

SECTION 30: SUSPENDED CEILINGS

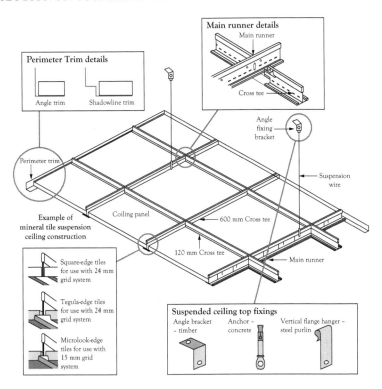

Figure 4.19 Typical suspended ceiling installation

Suspended ceilings take a variety of forms and can be installed for a number of reasons including:

- to improve sound insulation;
- to aid hygiene;

- to cover existing ceilings in a refurbishment project; and
- to provide a space to run M&E trunking and other equipment.

Suspended ceilings are generally formed by the installation of a lightweight metal grid suspended and shot fired to the structural soffit by wires of varying length (Figure 4.19). The grid then houses the tiles that form the suspended-ceiling finish; the grid may be concealed or form part of the finish. The main and cross tees are clipped together and around the perimeter of the ceiling angles are fixed to support the ceiling tiles at the perimeters. Suspended ceilings should be fitted with fire barriers (see Section 31 on how to prevent the spread of fire) and many have the facility to incorporate lighting.

Example

Armstrong suspended ceiling comprising 600 × 600 × 15 mm thick Dune tiles on and including 24 × 38 mm Prelude 24TLX grid, fixed to concrete with hangers, depth of suspension over 500 mm not exceeding 1 m. Cost per m².

Data

Armstrong tiles £80.40 per box 16 = 5.76 m²
3600 × 600 mm 24mm white main tee £4.30 each
600 × 24 mm white cross tee £1.00 each
Wire for suspension £10.00 per 50 m
Anchors for concrete – 10p each
Runner/cross-tee clips – 20p each
Assume a 3600 × 3600 mm grid (Figure 4.20), say 13 m²

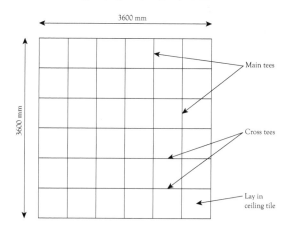

Figure 4.20 Suspended ceiling plan

£

Materials

36 ceiling tiles @ £80.40 per box of 16 = 2¼ boxes x £80.40	180.90
7 main tees @ £4.30	30.10
42 cross tees @ £1.00	42.00
49 anchors @ 0.10	4.90
49 clips @ 0.20	9.80
50 m wire @ £10.00 per 50 m	10.00
	277.70
Add waste 10%	27.77
	305.47

Labour

1 hour fitter per m² = 13 hours @ £23.16	301.08
	606.55
Add profit and overheads 15%	90.98
Cost per 13 m²	697.53
÷ 13 – cost per m²	£53.66

SECTION 31: INSULATION, FIRE STOPPING AND FIRE PROTECTION

Assume 10 m².

Example

200 mm thick mineral fibre insulation laid across joists, horizontal. Cost per m².

Data

2 m² pack of mineral fibre insulation £25.00

£

Materials

5 × 2 m² packs mineral fibre insulation @ £25.00 per pack	125.00
Add waste 5%	6.25
	131.25

Labour

0.3 hours carpenter/m² = 3 hours @ £23.16	69.48
	200.73

Add profit and overheads 15%	30.11
Cost per 10 m²	230.84
÷ 10 – cost per m²	£23.08

SECTION 32: FURNITURE, FITTINGS AND EQUIPMENT

Sanitary fittings

The following are labour constants for storing, assembling and fixing sanitary fittings:

Item	Unit	Labour (hours)
600 × 490 mm white vitreous china wash basin, pedestal, waste and pair 12 mm taps	Each	2.50
400 mm cantilever brackets	Pair	1.60
32 mm 'P' or 'S' trap	Each	0.50
40 mm 'P' or 'S' trap	Each	0.60
Low-level WC suite	Each	2.25
Cleaner's sink and waste	Each	1.00
1685 × 800 mm bath with mixer tap and 'P' trap including setting and levelling on feet plugging and screwing brackets to masonry and feet to concrete	Each	2.75
750 × 750 mm shower tray with waste, trap, mixer and shower head	Each	6.00
Bidet	Each	2.25
Single urinal with waste and automatic flushing cistern	Each	3.00

Example

600 × 490 mm white vitreous china wash basin and pedestal with one tap hole, plug and chain and 12 mm Accolade chromium plated basin mixer tap including plugging and screwing brackets to masonry and bedding pedestal to floor and basin in mastic. Cost per no.

Materials

	£
600 × 490 mm wash basin	108.90
Mixer tap	69.99
Waste, plug and chain	9.19
Jointing materials	2.00
Screws/plugs	1.50
	191.58

Add waste 5%		<u>9.58</u>
		201.16

Labour

2.50 hours @ £23.16		<u>57.90</u>
		259.06
Add profit and overheads 15%		<u>38.86</u>
Cost per unit		£<u>297.92</u>

Example

Close-coupled vitreous china WC unit with white plastic seat and cover, 9 litre cistern with dual flush and refill unit and P trap connector, including plugging and screwing to masonry and screwing pan to floor. Cost per no.

Materials

	£
Close-coupled WC	218.90
Toilet seat and cover	14.99
'P' trap connector	3.89
Screws and plugs	<u>2.00</u>
	239.78
Add waste 5%	<u>11.99</u>
	251.77

Labour

2.25 hours @ £23.16	<u>52.11</u>
	303.88
Add profit and overheads 15%	<u>45.58</u>
Cost per unit	£<u>349.46</u>

Example

355 × 540 × 400 mm vitreous china bidet with single lever mono bidet mixer tap plugged and screwed to concrete. Cost per no.

Materials

	£
Bidet	230.97
Taps	31.99

Jointing materials	2.00
Plugs and screws	1.50
	266.46
Add waste 5%	13.32
	279.78

Labour

2.25 hours @ £23.16	52.11
	331.89
Add profit and overheads 15%	49.78
Cost per unit	£381.67

SECTION 33: DRAINAGE ABOVE GROUND

Drainage above ground includes:

* rainwater installations; and
* foul drainage installations.

Unlike Section 38 Mechanical Services, fittings and ancillaries are measured and priced separately.

Rainwater gutters

Rainwater guttering is generally supplied in 2 m lengths and fixed with brackets at 1 m intervals.

Example

114 mm Hunter plastics Squareflo black straight rainwater gutter (ref. R114) with and including support brackets (ref. R395). Cost per m.

Data

114 mm Hunter plastics rainwater gutter £19.94 per 4 m length
Fixing brackets £0.95 each
Allow waste 5%
Assume 8 m

Materials

2 × 4 m rainwater gutter @ £19.94	£39.88
m ÷ 1 = 8 + 1 = 9 brackets @ £0.95 each	£8.55

	£48.43
Add waste 10%	<u>£4.84</u>
	£53.27

Labour

0.30 hour per m = 2.4 hours @ £23.16	<u>£55.58</u>
	£108.85
Add profit and overheads 15%	<u>£16.33</u>
Cost per 8 m	<u>£125.18</u>
÷ 8 – cost per m	<u>£15.65</u>

Sundry constants

Item	Outlet (hours/m)	Angle (hours/m)	Stop end (hours/m)
76 mm eaves gutter	0.25	0.25	0.20
114 mm eaves gutter	0.30	0.30	0.20
200 mm eaves gutter	0.40	0.40	0.25

Gutter ancillaries

Example

114 mm running outlet (ref. R376). Cost per no.

1 no. running outlet	£4.64
0.30 hours @ £23.16	<u>£6.95</u>
	£11.59
Add profit and overheads 15%	<u>£1.74</u>
Cost per no.	<u>£13.33</u>

Pipework

Rainwater pipe is generally supplied in 2 m lengths and fixed with brackets at 1 m intervals:

Item	Hours/m
56 mm rainwater pipe	0.30
74 mm rainwater pipe	0.35
160 mm rainwater pipe	0.50

Example

110 mm Hunter plastics rainwater pipe (ref. S506) with push-fit socket joints and stand off brackets (ref. S219) at 2 m centres plugged to masonry. Cost per m.

Data

110 mm Hunter plastics rainwater pipe £42.18 per 4 m length
Fixing brackets £3.03 each
Allow 5% waste
Assume 8 m

Materials

2 x 4 m rainwater pipe @ £42.18	£84.36
9 no. brackets @ £3.03 each	£27.27
	£111.63
Add waste 5%	£5.58

Labour

0.30 hour per m = 2.4 hours @ £23.16	£55.58
	£172.79
Add profit and overheads 15%	£25.92
Cost per 8 m	£198.71
÷ 8 – cost per m	£24.84

Pipework fittings

SUNDRY CONSTANTS

Item	Shoe (hours/no.)	Offset (hours/no.)	Single branch (hours/no.)
56 mm rainwater pipe	0.30	0.35	0.35
74 mm rainwater pipe	0.35	0.40	0.40
160 mm rainwater pipe	0.50	0.55	0.55

Example

56 mm rainwater shoe (ref. S322). Cost per no.	
1 no. rainwater shoe	£18.39
0.30 hours @ £23.16	£6.95

	£25.34
Add profit and overheads 15%	£3.80
Cost per no.	£29.14

Foul drainage installation

Foul drainage installation includes the pipework and fittings associated with sanitary appliances and also includes all the necessary fittings to make connection to the sanitary appliances.

Sundry constants

Pipe diameter	Pipe (hours/m)	Bend (hours/m)	Single junction (hours/no.)	Offsets (hours/no.)
19 mm	0.30	0.25	0.30	0.25
40 mm	0.30	0.25	0.40	0.25
82 mm	0.35	0.30	0.45	0.30
110 mm	0.35	0.35	0.50	0.35
160 mm	0.50	0.40	0.50	0.40

Example

110 mm straight pipes solvent weld PCV-U (ref. SS300) with and including pipe bracket (ref. US221) plugged and screwed to masonry. Cost per m.

Data

110 mm pipe £11.38 per 3 m delivered to site
Socket bracket £1.19 each
Coupler £4.08 each
Allow 5% waste
Assume 3 m

Materials

3 m waste pipe @ £11.38 per 4 m	£11.38
1 no. coupler	£4.08
2 no. socket brackets @ £1.19	£2.39
	£17.85
Add waste 5%	£0.89
	£18.74

Labour

0.35 × 3 = 1.05 hours @ £23.16	£24.32
	£43.06
Add profit and overheads 15%	£6.46
Cost per 3 m	£49.52
÷ 3 – cost per m	£16.51

Maximum support centres for pipes

	Centres (m)	
Pipe size	Vertical	Horizontal
82 mm	2.0	1.0
110 mm	2.0	1.0
160 mm	2.0	1.2

SECTION 34: DRAINAGE BELOW GROUND

Excavation

Excavation to drain trenches can be based on the labour constants given for surface trenches except that allowance has to be made in grading the bottoms of the trench to the correct gradient. Drainage trenches are measured in linear meters and are deemed to include all necessary earthwork support, consolidation of bottom of trench, trimming excavation, filling and compaction of general filling materials and removal of surplus excavated materials (Figure 4.21). The following is a guide to the width of a drainage trench based on pipes up to 200 mm diameter although the actual width is at the discretion of the contractor:

Depth of trench	Width
Average depth up to 1 m	500 mm
Average depth 1 to 3 m	750 mm
Average depth over 3 m	1,000 mm

For pipes over 200 mm diameter, the width of the trench should be increased by the additional diameter of the pipe over 200 mm.

Example

The quantities of excavation, earthwork support, etc. are calculated for each category of pipe and depth. Once calculated, they can be applied to any project.

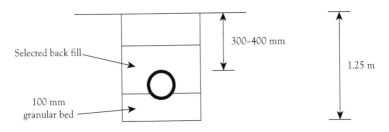

Figure 4.21 Typical section through pipe trench

The following items are deemed to be included with drain runs:

- earthwork support;
- consolidation of trench bottoms;
- trimming excavations;
- filling with and compaction of general filling materials;
- disposal of surplus excavated materials;
- disposal of water;
- building in the end of pipes; and
- bedding and pointing the length of pipe with manhole walls.

Backfilling with material arising from the excavations is deemed to be included; however, backfilling in selected material must be identified and described. The exact nature of the backfilling required will depend on the pipes being used.

Pipe beds, haunchings and surrounds

Once laid in the trench, protection to the drain pipes is provided by beds under the pipe, haunchings to the sides of the pipe or complete surrounds to the pipes depending on the location of the pipes and the specification. If concrete is used, any formwork is deemed to be included in the item.

Drain pipes

The traditional material for below-ground drainage pipes and fittings is clay; however, this material has now been largely replaced with UPVC (plastic) pipes. Clay pipes, now almost all of which are referred to as vitrified pipes, are still used in certain situations and are manufactured in 600 mm lengths. The big advantage of UPVC pipes is that they weigh considerably less than vitrified clay pipes and come in 3 m lengths, making handling, laying and jointing easier and quicker. Pipes of both materials are jointed with flexible push-fit joints that helps to prevent failure.

Example

110 mm Floplast pipe laid in trenches average 1.5 m deep with coupling sockets joints on 100 mm granular bed and surrounded with 150 mm thick selected excavated material. Cost per m.

Excavation

Assume 10 linear metres

Excavation:

	10.00		
	0.75		
	<u>1.25</u>	9.38 m³ @ £13.34	125.13

Earthwork support:

	2/10.00		
	<u>1.25</u>	25.00 m² @ £7.25	181.25

Compaction to bottom of trench:

	10.00		
	<u>0.75</u>	7.50 m² @ £1.20	9.00

General backfill:

	10.00		
	0.75		
	<u>0.80</u>	6.00 m³ @ £2.00	12.00

Remove surplus:

	10.00		
	0.75		
	<u>0.45</u>	3.38 m³ @ £4.98	<u>16.83</u>

Cost per 10 m   £<u>344.21</u>

÷ 10 – cost per m (excavation)   £34.42

Drain pipes

110 mm Floplast pipes £19.98 per 6 m
Coupling sockets £3.68 each
Assume 30 m

Materials

5 × 6 m Floplast pipes @ £19.98 each	99.90
10 coupling sockets @ £3.68	<u>36.80</u>
	136.70

Labour

Laying and jointing pipes – 0.1 hours per m	
3 hours @ £23.16	69.48
Cost per 30 m	206.18
÷ 30 – cost per m (pipework)	£6.87

Bed and surround to pipe

Assume 10 linear metres
100 mm granular bed

Materials

$10.00 \times 0.75 \times 0.10 = 0.75$ m³ selected granular bed @ £10.00 per m³	£7.50
$10.00 \times 0.75 \times 0.30 = 2.25$ m³ selected excavated material surround @ £5.00 per m³	£11.25

Labour

10 m bed and surround pipes @ 0.2 hour per m	
2 hours @ £23.16	£46.32
Cost per 10 m	£65.07
÷ 10 – cost per m (bed and surround)	£6.51

Summary

Excavation	£34.42
Pipework	£6.87
Bed and surround	£6.51
	£47.80
Add profit and overheads 15%	£7.17
Total (per m)	£54.97

Manholes and inspection chambers

Traditionally, the construction of manholes was a labour-intensive operation involving many trades. Although still constructed from engineering bricks, it is more common to use preformed manholes made from precast concrete, plastic and clay. The labour constants for work to manholes can be taken from the respective trades with the addition of about 25% to take account of working in confined spaces. Inspection chambers are essentially the same as manholes but are usually shallower.

NRM2 states that all manholes, whether precast or in-situ are to be enumerated and therefore it is necessary for the estimator to build up rates from details on the drawings.

SECTION 35: SITE WORKS

This section includes work to road and path pavings, hard landscaping and sports surfacing and includes trades previously discussed in this chapter such as excavation and concrete work (see Figures 4.22 and 4.23).

Example

255 × 125 mm half-batter precast concrete kerb on and including 300 × 150 mm plain in-situ concrete class D foundation and haunch.

Note that NRM2 states that excavation and disposal is deemed to be included with this item. In practice, these items would be carried out during the reduce level excavation for the road sub-base.

Data

255 × 125 × 900 mm long half-batter kerb £11.45 each
Assume 9 linear metres

Materials

10 half-batter kerbs @ £11.45	114.50
Class D concrete bed and haunch (Figure 4.23)	

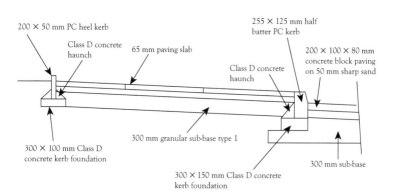

Figure 4.22 Section through estate road

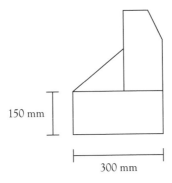

150 mm

300 mm

Figure 4.23 Bed and haunch to kerb

Bed:
$9.00 \times 0.30 \times 0.15$ $= 0.41$ m^3

Haunch:
$\frac{1}{2} \times 9.00 \times 0.18 \times 0.18 = \underline{0.15}$ m^3
                 0.56 m^3 per 9 m @ £90.44/m^3 50.65

Labour

0.2 hours per linear meter craftsman and labourer
1.8 hours @ £38.91   <u>70.04</u>
                                                   235.19
Add profit & overheads 15%   <u>35.28</u>
Cost per 9 m   <u>270.47</u>
÷ 9 cost per m   <u>£30.05</u>

SECTION 36: FENCING

Fencing is required to be measured in linear metres by NRM2 and when estimating the cost of fencing the following items have to be included in the rate:

- excavation and concreting or driving of posts;
- fence posts; and
- fencing: post and rail, post and wire or interwoven.

Gates are enumerated, as is ironmongery. Allow 5% waste.

Sundry constants

Item	Unit	Hours
Excavate post hole, including part backfill and remove surplus	m^3	3.00 (unskilled)
Concrete in post hole	m^3	2.50 (unskilled)
100 × 100 m strut to concrete post	no.	0.60 (unskilled)
Hanging five-bar gate 3.5 m wide on field-gate hinges	no.	2.50 (skilled)
Fix only field-gate fastener	no.	1.00 (skilled)

Example

1200 mm high impregnated softwood morticed post and three-rail fence comprising 125 × 75 × 1800 mm long posts at 2.85 m centres, 100 × 38 × 1500 mm prick posts and three no. 100 × 38 mm rails. Cost per 3 m bay.

Data

1800 mm three morticed impregnated softwood fence post £17.10 each
100 × 38 × 3000 mm long impregnated softwood rails £10.00 each
100 × 38 × 1500 mm prick post the intermediate post£6.20 each
50 mm galvanised 2.65 gauge wire nails £4.50/kg

Other items deemed to be included but nevertheless have to be included in the rate include excavation, disposal and concrete for post bases

Allow 11 kg of 100 mm galvanised wire nails for 200 mm of three-rail fence and 14 kgs for 200 mm of four-rail fence

Assume 2.85 × 2 = 5.7m = 3 no. posts, 2 no. prick posts and 6 no. rails

Posts will have concrete bases, but prick posts will be driven into the ground

Labour £

3 no. 400 × 400 × 600 mm deep post hole = 0.1 m^3

Excavating post holes:
    3 × 0.1 = 0.3 m^3 × 3 hours = 0.9
    0.9 hours unskilled @ £15.75   14.18

Filling concrete into post holes:
    3 × 0.07 m^3 × 0.21 × 2.5 hours = 0.53
    0.53 hours unskilled @ £15.75   8.35

Setting post in concrete:
    3 × 0.3 hours per post @ £23.16   20.85

Fixing rails and prick post:
    3 × 2 rails = 6 rails @ 0.1 hour per rail =   0.60
    2 × 1 prick posts = 2 posts @ 0.1 hour/post = <u>0.20</u>
                                    0.80
    0.80 hours @ £23.16   <u>18.53</u>
                                          61.91   61.91

Materials

3 no. posts @ £17.10	51.30		
2 no. prick posts @ £6.20	12.40		
6 no. rails @ £10.00	<u>60.00</u>		
	£123.70	123.70	
Nails		<u>8.50</u>	
		132.2	
Add waste 5%		<u>6.61</u>	
		138.81	<u>138.81</u>
			200.72
Add profit and overheads 15%			<u>30.11</u>
Cost per 3 m bay			<u>£230.83</u>

SECTION 37: SOFT LANDSCAPING

Soft landscaping includes the planting and maintenance of trees, shrubs, plants and bulbs, as well as seeding and turfing.

SECTION 38: MECHANICAL SERVICES

Mechanical engineering operatives have their own separate wage negotiation structure and dayworks charges – see Chapter 5.

    The negotiation of the hourly rate for plumbing operatives is carried out by the Joint Industry Board for Plumbing Mechanical Engineering Services in England and Wales (JIB). Scotland and Northern Ireland have their own board. In 2017, a three-year pay deal was agreed. The following is based on 2019 rates and, as with BATJIC, there is a wide variety of pay scales reflecting the levels of competency and skill. As with BATJIC, tool allowances are now incorporated into the rates.

    The all-in-rate for plumbing technician operatives per hour is calculated as follows:

	Technician operative £	£
Standard JIB wage rate 1,687.5 hours	17.19	29,008.13
Overtime (basic rate up to 8 pm) 225 hours	17.19	3,867.75
Plumbers welding supplement 1,912.5 hours	0.56	1,071.00
Gross earnings		33,946.88
Add		
Employer's contribution to:		
Holiday credit (to provide for 31 days (241 hours, 32 from 2020))		4,142.79
Plumbing and Mechanical Services Industry Pension Scheme (7.5% of gross earnings)		2,446.77
National Insurance (13.80% payable on gross earnings above threshold)		3,522.18
		44,058.62
Severance pay and sundry costs plus 1.5%		660.88
Employer's liability and third-party insurance plus 2.0%		881.17
Total cost per annum		£45,600.67
÷ 1,912.5 – cost per hour		£23.84

Note: The normal working week is 37.5 hours.

Allowances

Daily travel time allowances

Note that these allowances are payable in addition to fares:

Over (miles)	Not over (miles)	All operatives
20	30	£4.86
30	40	£11.34
40	50	£12.96

In addition to the above, the cheapest available return fares will be payable.

Subsistence allowance

This is subject to PAYE deductions for London only of £5.58 per day.

Lodging allowance

This is paid at £38.60 per night.

Responsibility/incentive pay allowance

Employers may, in consultation with employees, enhance the basic rates of pay by the payment of an additional amount, as per the bands below, where it is agreed that their work involves extra responsibility, productivity or flexibility:

Band	Additional hourly rate
1	1p to 31p
2	32p to 52p
3	53p to 78p
4	79p to £1.02

Perhaps to reflect the fact that Scotland and Northern Ireland have more sole proprietors, the Scottish & Northern Ireland Joint Industry Board for the Plumbing Industry agreed that the minimum hourly rate to be paid to labour-only self-employed operatives should be £15.61.

Mechanical services covers a number of types of installations and for the purposes of this pocket book, the following will be considered:

- rainwater installations – included in Section 33: drainage above ground;
- disposal installations – included in Section 33: drainage above ground;
- domestic water installations.

Although it should be noted that NRM2 does not contain a specific requirement for these installations to be categorised as shown above, NRM1 does use these categories for estimating and cost-planning purposes.

It should be noted that, unlike Section 33: drainage above ground, NRM2 gives two alternatives for the measurement of pipework and fittings:

- default alternative – pipes are measured inclusive of fittings; and
- alternative 1 – fittings are enumerated and measured separately.

The estimator should note that whichever of these alternatives has been used, in the preparation of the bill of quantities or work package the cost of fittings must still be included in the build-up of rates.

Hot- and cold-water installation

Allow one pipe clip per linear metre of pipe and one straight coupling per three linear metres of pipe as copper tube is manufactured in 3 m lengths. Copper tube

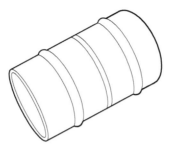

Figure 4.24 Capillary coupling for copper tube

can either be jointed with capillary joints or compression joints, although capillary joints are used more frequently since this method is cheaper and quicker than compression joints (Figure 4.24).

A capillary joint consists of a copper sleeve with socket outlets into which the pipe ends are soldered. It is neater and smaller than a compression joint and forms a strong connection that will not easily pull apart. Because a capillary fitting costs much less than a compression fitting, they are often used when a number of joints have to be made, such as when installing a new central heating system. They are also useful in cramped spaces, where it would be impossible to use wrenches to tighten a compression joint. The most common type of capillary joint has a ring of solder pre-loaded into the sleeve. It is known as an integral ring or 'Yorkshire' fitting (the name of a leading brand).

Compression joints (Figure 4.25) are more expensive than capillary fittings. They are made up of a central body of brass or gun metal with a cap-nut at the end and an olive to help give a watertight seal. When the cap-nut is rotated clockwise, it squeezes the olive tightly between the pipe end and the casing. Getting a watertight seal depends upon the ends of the pipe having been well prepared so they will butt

Figure 4.25 Compression coupling for copper tube

up exactly to the pipe stop in the casing without any gaps. This forms a seal and ensures that the pipe is parallel to the movement of the rotating cap-nut. The cap-nut applies an even pressure to the olive so that it does not buckle under the strain of being tightened. Joint paste is sometimes used to help ensure against leakage.

Add 2.5% waste generally and 5% waste for work in short lengths.

Sundry labour constants: light gauge copper tube

Diameter	Hours per linear metre
15 mm	0.20
22 mm	0.20
28 mm	0.25
42 mm	0.30

Example

15 mm copper tube to BS EN 1057 Table X with capillary fittings, fixed with pipe clips to masonry. Cost per m.

Data

15 mm copper tube £6.19 per 2 m
Couplings £4.09 for 10
Pipe clips £2.29 per 100
Assume 10 m

Materials

10 m of 15 m diameter copper tube @ £6.19 per 2 m	£30.95
Couplings 10 ÷ 3 = 3 + 1 +1 = 5 couplings @ £0.41	£2.05
Pipe clips 10 + 1 = 11 pipe clips @ £0.02	£0.22
	£33.22
Add waste 2.5%	£0.83

Labour

10 m copper tube @ 0.20 hours/m = 2 hours @ £23.84	£47.68
	£81.73
Add profit and overheads 15%	£12.26
Cost per 10 m	£93.99
÷ 10 – cost per m	£9.40

Capillary fittings: light gauge copper tube – hours per item

Diameter	Elbow	Tee	Made bend
15 mm	0.20 (Figure 4.26)	0.30	0.20
22 mm	0.25	0.35 (Figure 4.27)	0.20
28 mm	0.25	0.35	0.35
42 mm	0.30	0.40	0.45

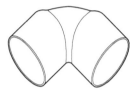

Figure 4.26 15 mm copper capillary elbow

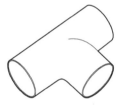

Figure 4.27 22 mm copper capillary tee

Example

Capillary fittings and jointing to light gauge copper tube. 15 mm diameter elbow – cost per no.

Data

15 mm diameter elbow £4.99 per 10

Materials

10 no. 15 mm elbows @ £4.99	4.99
Add waste 5%	0.25
	5.24

Labour

10 elbows @ 0.2 hours each = 2 hours plumber @ £23.84	<u>47.68</u>
	52.92
Add profit and overheads 15%	<u>7.94</u>
Cost per 10	<u>60.86</u>
÷ 10 cost per number	**<u>£6.09</u>**

Example

22 mm diameter equal tee. Cost per no.

Data

22 mm diameter equal tee £1.00

Materials

Assume 10	
10 no. 22 mm tees @ 1.00	10.00
Add waste 5%	<u>0.50</u>
	10.50

Labour

10 tees @ 0.35 hours each = 3.5 hours plumber @ £23.84	<u>83.44</u>
	93.94
Add profit and overheads 15%	<u>11.75</u>
Cost per 10	<u>105.69</u>
÷ 10 – cost per no.	<u>£10.57</u>

In addition to the fittings described above, hot and cold water systems also require a number of fittings (ancillaries) to control the flow of water and facilitate maintenance by isolating parts of the system (Figure 4.28a–c).

SUNDRY LABOUR CONSTANTS

Stop valve	Labour (plumber, hours)
15 mm diameter	0.60
22 mm diameter	0.70
28 mm diameter	0.85

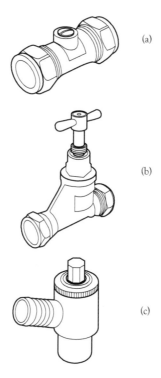

(a)

(b)

(c)

Figure 4.28 (a, b) Stop valves. (c) Drain cock. Both isolating valve and stop valve have compression joints to copper pipe

Example

15 mm brass stop valve with compression joints to copper.

Data

15 mm brass stop valve £3.76 each
Assume 10 valves

Materials

10 valves @ 3.76 each	37.60
Add waste 5%	<u>1.88</u>
	39.48

Labour

10 valves @ 0.60 hours per valve = 6 hours @ £23.84	<u>143.04</u>
	182.52
Add profit and overheads 15%	<u>27.38</u>
Cost per 10 valves	<u>209.09</u>
÷ 10 cost per valve	£20.91

Polybutylene hot and cold water piping and fittings

In addition to traditional copper pipes and fitting, several manufacturers now supply polybutylene piping suitable for the full range of water services: hot and cold supply and central heating in a range of sizes from 10 mm to 35 mm diameter that is fully compatible with copper. Special push-fit fittings are also made as part of each manufacturer's system. Generally, the wall of the pipe requires support at each fitting and this is given by the insertion of a shouldered liner into the end of the pipe before pushing it home (Figure 4.29).

Any type of pipe clip that can be used with copper pipe can be used with polybutylene pipe but those comprising a metal band are best avoided as cyclical thermal movement of the pipe in the clip can result in the metal cutting through the pipe wall. Snap-in plastic clips or plastic saddles are preferred. The pipe is available in cut lengths generally of 3, 4 and 6 m and in coils of up to 100 m. Whichever length is supplied, the pipe has the same flexibility and this flexibility is promoted by the manufacturers as of particular advantage in running the pipe in confined spaces or through multiple timbers as in joisted floors or timber partitions.

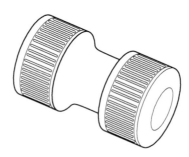

Figure 4.29 Straight coupling for polybutylene pipe

Sundry constants

Item	Hours (plumber per m)
15 mm pipe screwed to softwood	0.25
22 mm ditto	0.30
28 mm ditto	0.35

Fittings	Hours (plumber, each)
15 mm elbow	0.05
22 mm ditto	0.07
15 mm tee	0.08
28 mm tee	0.15

Polybutylene pipe has

- low thermal conductivity coupled with flexibility enabling it to cope with low temperatures;
- high impact resistance;
- anti-corrosive properties;
- flexibility; and
- an oxygen barrier added that reduces ingress of it through the wall.

Advantages of polybutylene fittings:

- quick push-fit system;
- no tools required;
- will expand and contract with the pipe;
- fittings are demountable; and
- non-toxic.

Applications of polybutylene:

- indirect/direct mains-fed cold-water services;
- vented/unvented hot-water systems; and
- underfloor heating systems.

SECTION 39: ELECTRICAL SERVICES

In Chapter 3 Resources, the calculation of the all-in hourly rate for building opera-tives was illustrated; electrical contractors and sub-contractors have their own separate mechanism for determining conditions of work and rates of pay. The body that looks after the interests of electrical services contractors/sub-contractors is the Joint Industry Board for the Electrical Contracting Industry (JIB). The JIB has its own National Working Rules and Industrial Determinations and governs and controls the conditions for electrical, instrumentation and control engineering, data and communications transmission work, its installation, maintenance and its dismantling and other ancillary activities. The JIB replaced existing bodies and its authority came into effect in respect of work performed on and after Tuesday 2 January 2001.

JIB National Standard Rates

One obvious difference between the BATJIC hourly wage rates and the JIB is the greater number of operatives grades (Table 4.1).

Operatives working in London, as defined in the JIB working rules, receive an approximate 15% in addition to the above rates.

Other items that should be noted are:

Working time

The standard working time is 37.5 hours per week.

Overtime

Overtime is paid at time and a half. All hours worked on Saturday in excess of the first six or worked after 3.00 pm Saturday, whichever comes first, and before the normal starting time on Monday shall be paid at double time. Overtime premium payments shall be calculated on the appropriate standard rate of pay.

Call-outs

For emergency call-out(s) when an operative, having returned home after his normal finishing time, is called upon to return to work before his next normal starting time, he shall be paid at time and a half for all hours worked home-to-home.

Table 4.1 JIB National Standard Rates

Grade	Transport provided	Own transport
Technician	£17.06	£17.92
Approved electrician	£15.08	£15.92
Electrician	£13.81	£14.68
Electrical improver	£13.14	£13.95
Labourer	£10.97	£11.79
Senior graded elec. trainee	£12.42	£13.23
Adult trainee	£10.97	£11.79

As with overtime, all hours worked on Saturday in excess of the first six or worked after 3.00 pm Saturday, whichever comes first, and before normal starting time on Monday shall be paid at double time.

In addition, an operative shall be paid a call-out allowance of:

- Single call-out: £20.00 for a graded operative or £7.50 for an apprentice.
- Second and subsequent call-out(s): In the event an employee is, having returned home after call-out(s), called again, a further allowance of: £10.00 for a graded operative or £3.50 for an apprentice shall be paid in respect of this second and, at the same rates, for each subsequent call-out prior to the next normal starting time.

Traveling time and allowances

Straight line distance from job to shop	Mileage allowance	Mileage rate
Up to 15 miles each way	Nil	Nil
Over 15 miles each way	22p per mile	12p per mile

Lodging allowances and retention payments

From and including 2 January 2019 the Lodging Allowance is:

Lodging Allowance	£39.69 per night
Retention payments to a maximum of	£13.06 per night (£91.42 per week)
Weekend retention fees	£39.69

Annual holidays

Operatives are currently entitled to payment for 22 days (23 days from 7 January 2019; 24 days from 6 January 2020) annual holidays plus 8 days public holidays as determined from time to time by the JIB National Board.

Sickness with pay

	Others*	Electrician	Approved electrician	Technician
Weeks 1–2	Nil	Nil	Nil	Nil
Weeks 3–24	£150.00	£160.00	£170.00	£180.00
Weeks 25–52	£75.00	£80.00	£85.00	£90.00

* 'Others' covers labourer, adult trainee, senior graded electrical trainee and electrical improver grades.

Bereavement leave

Employers are required to give sympathetic consideration to requests from operatives for bereavement leave in the event of the death of a close relative (e.g. child, spouse, partner, parent). When such bereavement leave is granted, employers will pay the operative concerned for up to three normal working days at their basic hourly rate.

Disciplinary and dismissal procedure

There are lengthy disciplinary and dismissal procedures together with grievances and mediation procedures in the case of disputes.

As a callow youth, I bought an estimating book to help me pass the RICS professional examinations; whereas most trades were comprehensively covered, when it came to electrical installations the book stated 'this section is highly specialised and is not dealt with in this book'. The perception that electrical work is something like a black art still persists in the industry although, in practice, the basic approach is very much the same as for other trades.

What is true is that for the majority of small–medium size projects, there will be little or no detailed drawn information and therefore it may be necessary for the estimator to quantify what is required prior to pricing. It should be remembered however at all times that the responsibility for achieving compliance with the requirements of Part P Building Regulations, in the case of domestic properties, rests with the person carrying out the work. That 'person' may be, for example, a developer, a main contractor or a sub-contractor, or specialist firm directly engaged by the client.

Another feature of M&E installations is the use of performance specifications. A performance specification is a non-prescriptive document that sets out the level of performance required from the completed installation without going into detail about what specific materials or equipment should be used. The finished installation must in all respects comply with the necessary regulations; the contractor or sub-contractor has a free rein to decide how the output parameters will be achieved. Figures 4.30 and 4.31 illustrate a typical layout for the electrical installation to a domestic dwelling.

Figure 4.30 illustrates a domestic typical layout at the point where the electricity distributor brings the supply into the building. The main fuse (cut out) and the meter are the responsibility of the distributor and will be supplied and installed in a safe location by that company, and therefore they are not usually required to be measured and priced as part of this section. From the meter the supply then goes to the consumer service unit (often called the fuse box although fuses were replaced with miniature circuit brakers (MCBs) many years ago), which contains a number of miniature circuit breakers of varying ratings from which the lighting, ring mains and appliance circuits are run; it is from this point, the consumer service unit, that provision has to be made in the bill of quantities or work package. Cables are generally concealed within stud partitions, ceilings and roof voids or, in the case of

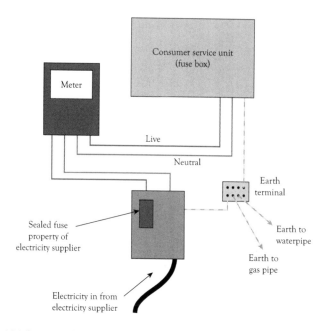

Figure 4.30 Incoming electricity supply

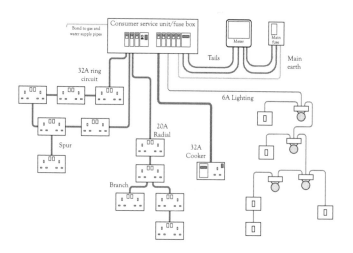

Figure 4.31 Typical electrical layout for domestic scale project

masonry walls and partitions, in a chase subsequently covered by finishes. Surface-mounted cables have to be protected from damage by running them in trunking or conduit. Figure 4.31 has been simplified for illustrative purposes but could also commonly include circuits for central heating, alarm systems, electric showers and outdoor sockets/lighting, all of which require their own MCB. Any chasing in brick-work or blockwork should be included in Section 41: builder's work in connection with mechanical, electrical and transportation installations.

Calculation of electrical operatives' all-in hourly rate

Hours available for work in accordance with the JIB Working Rule Agreement:

- 37.5 hour week; and
- 30 days holiday a year – 8 public and 22 annual.

Summer period

Number of weeks	30	
Weekly hours	37.5	
Total hours		1,125.00
Less holidays		
Annual – 14 days		(112.50)
Public – 4 days		(30.00)

Winter period

Number of weeks	22	
Weekly hours	37.5	
Total hours		825
Less holidays		
Annual – 8 days		(60.00)
Public – 4 days		(30.00)
Sickness	8	(60.00)
Total hours for payment		1,657.50

The all-in-rate for electrical operatives per hour is calculated as follows:

	Technician		Labourer	
	£	£	£	£
Standard JIB wage rate 1,657.5 hours	17.06	28,276.95	10.97	18,182.78
Overtime (basic rate) 225 hours	17.06	3,889.68	10.97	2,468.25
Overtime	23.07		14.84	
		32,166.63		20,651.03

Incentive scheme 5%	1,608.33	1,032.55
Gross earnings	33,774.99	21,683.58
Add:		
BlueSky pension scheme of gross earnings 2.5%	844.38	542.09
JIB combined benefit scheme 52 weeks × £15.15	787.80	787.80
National Insurance – 13.8% payable on gross earnings above threshold*	3,498.44	1,829.82
	38,905.61	24,843.29
Employer's liability and third-party insurance plus 2.0%	778.11	496.87
Total cost per annum	**£39,683.72**	**£25,340.16**
÷ 1,657.5 – cost per hour	**£23.94**	**£15.29**

* Variable percentage – check with HMRC.

A further addition is the traveling time and allowance rate, which will vary from job to job depending on the distance of the project from the shop.

Example

Primary equipment. Safeclip two-way 63-amp high-rupturing capacity distribution fuse board with hinged metal enclosure complete with two high-rupturing capacity (HRC) fully shrouded 63-amp fuse holders and links including plugging and screwing to masonry. Cost per no.

Data

Hager distribution board £75.50 each
63-amp HRC fuses £3.05 each

Materials

1 no. Hager distribution board	75.50
2 no. HRC links @ £3.05 each	6.10
	81.60
Add waste 5%	4.08
	85.68

Labour

1 hour technician @ £23.94	23.94
	109.62
Add profit and overheads 15%	16.44
Cost per unit	**£126.06**

Example

16-way consumer service unit comprising metal enclosure with hinged cover plate complete with:

- 60-amp double-pole isolating switch;
- seven 10-amp MCBs;
- four 16-amp MCBs;
- two 32-amp MCBs; and
- three blank plates.

Plugged and screwed to masonry. Cost per no.

Data

MK 16-way consumer service unit £148.28

(Note some consumer service units do not include MCBs and these must be bought separately.)

Materials

MK 16-way consumer service unit	148.28
7 no. 10-amp MCBs @ £3.31	23.17
4 no. 16-amp MCBs @ £3.49	13.96
2 no. 32-amp MCBs @ £3.60	7.20
	192.61
Add sundries (cable and clips, etc.) 5%	9.63
	202.24
Add waste 5%	10.11
	212.35

Labour

2 hours technician @ £23.94	47.88
	260.23
Add profit and overheads 15%	39.04
Cost per unit	**£299.27**

Example

Terminal equipment and fittings. MK Superswitch surface mounted one-gang 13-amp switched socket outlet (ref. MKSW1) including plugging and screwing to masonry. Cost per no.

Data

MK Superswitch £10.00 per pack of 5.

Materials

Assume 5 switches	
5 no. switches @ £10.00 for 5	10.00
Add sundries, say 5%	<u>0.50</u>
	10.50
Add waste 5%	<u>0.53</u>
	11.03

Labour

5 no. switches @ 0.20 hour per switch	
1 hour technician @ £23.94	<u>23.94</u>
	34.97
Add profit and overheads 15%	<u>5.25</u>
	40.22
÷ 10 – cost per no.	<u>£4.02</u>

Cable containment

Cables can be contained in trunking or ducting and NRM2 allows for two methods of billing:

- generally all cable containment to be measured inclusive of all fittings, i.e. joint boxes, connectors, flanges, bends, tees, junctions, reducers, spigots, fire barriers and the like; or
- alternative 1 (39.4) allows for joint boxes, connectors, flanges, bends, tees, sets, junctions, reducers, spigots, fire barriers and the like to be measured separately.

If alternative 1 is used and fittings are measured separately, those shown in Figures 4.32 to 4.34 are common.

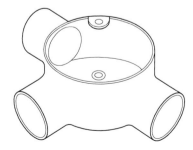

Figure 4.32 Three-way conduit box

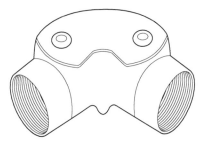

Figure 4.33 Inspection elbow

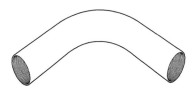

Figure 4.34 Bend

Example

20 mm heavy gauge enamelled steel conduit with and including nylon draw wire fixed to masonry. Cost per linear metre.

Data

20 mm Conduit £8.20 per 3 m
Spacer bar saddles £0.60 per pack of 2
20 m draw wire £5.80 each

Materials

Assume 3 linear metres	
3 m of 20 mm conduit	8.20
Add for fittings, say 10%	0.82
20 mm draw wire £5.80	
Assume 10 uses ÷ 10	0.58
Spacer bar saddles – taken @ 1 m centres	
Say, 4 saddles @ £0.60 per 2	1.20
	10.80

Labour

0.30 hours per linear metre technician
0.90 hours @ £23.94   21.55
                                                               32.35
Add profit and overheads 15%   4.67
Cost per 3 m   37.02
÷ 3 – cost per m   £12.34

Example

Cables. 2.5 mm^2 PVC insulated and sheathed two core cable fixed to masonry. Cost per m.

Data

2.5 mm^2 PVC cable £63.49 per 100 m
Cable clips £2.99 per 100

Materials

Assume 100 m
100 m 2.5 mm^2 cable @ £63.49 per 100 m   63.49
100 cable clips @ 1 m centres   2.99
                                                               66.48
Add waste 5%   3.32
                                                               69.80

Labour

Hours per metre technician =
10 hours @ £23.94 per hour   239.40
                                                             309.20
Add profit and overheads 15%   46.38
Cost per 100 m   355.58
÷ 100 – cost per m   £3.56

Final circuits

Final circuits are enumerated as the exact layout is left to the discretion of the electrical contractor. When pricing the circuit, the estimator will have to study the drawings and make an assessment each type of cable and conduit to allow in each circuit, based on experience.

Example

Final circuit surfaced mounted 2.5 mm PVC insulated single core cables drawn into heavy gauged black enamelled steel conduit, plugged and screwed to masonry with and including heavy gauged spacer bar saddles at 1 m centres from consumer control unit serving rooms 21 and 22 with one one-gang switched socket outlet and three two-gang switched socket outlets. Cost per no.

Data

2.5 mm PVC insulated single core cable £63.49 per 100 m
22 mm conduit £8.20 per 3 m
Draw tape £5.80 each
Spacer bars £0.60 per pack of 2
One-gang switched socket outlet £10.00 per pack of 5
Two-gang switched socket outlet £17.50 per pack of 5

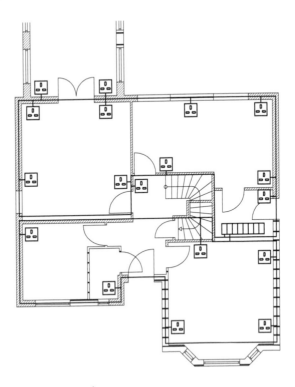

Figure 4.35 Final circuit example

Single knockout box £0.36 each
Double knockout box £0.42 each

Note: the actual lengths of cabling, etc. and the number of outlets will be ascertained with reference to the project drawings. The number and position of switches, sockets and luminaries will be indicated on the drawings using the symbols, although it should be noted that the symbols can alter from engineer to engineer (Figure 4.35).

	£
Materials	
(Note: lengths calculated from drawings)	
25 m single-core cable @ £63.49 per 100 m	15.88
25m 22mm conduit @ £8.20 per 3m $\dfrac{25}{3} = 8.33 \times £8.20$	68.31
Fittings, say 10%	6.83
Draw tape	5.80
Spacer bar saddles 26 @ £0.60 per 2	7.20
One-gang switched socket outlet	2.00
One single knockout box	0.36
3 × two-gang switched socket outlets	7.50
3 × double knockout boxes	<u>1.26</u>
	115.14
Add waste 5%	<u>7.76</u>
	122.90
Labour	
One-gang socket outlet 2 hours @ £23.94	47.88
3 × two-gang socket outlet	
3 × 2.5 hours = 7.5 hours @ £23.94	<u>179.55</u>
	350.33
Add profit and overheads 15%	<u>52.55</u>
Cost per no.	<u>£402.88</u>

SECTION 40: TRANSPORTATION

Transportation is a work section new to NRM2 and appears to be a catch-all section to accommodate allowing contractors and sub-contractors the opportunity to price for transport costs associated with large items manufactured off-site for incorporation into the works.

SECTION 41: BUILDER'S WORK IN CONNECTION WITH MECHANICAL, ELECTRICAL AND TRANSPORTATION INSTALLATIONS

Builder's work in connection includes all the general builder's items such as:

- marking position of holes, mortices and chases in the structure;
- pipe and duct sleeves;
- cutting holes through existing structures; and
- testing and commissioning.

Example

Cutting chase in existing brickwork for 22mm diameter pipe. Cost per m. Assume 10 m.

	£

Labour

Labourer cuts 1 m of chase in 0.15 hours using a power hammer	
1.5 hours @ £15.75	23.63
Add profit and overheads 15%	3.55
Cost per 10 m	27.18
÷ 10 – cost per m	£2.72

Example

Cutting 20 mm diameter hole through one-brick wall. Cost per no.

	£

Labour

0.50 hours @ £15.75	7.88
Add profit and overheads 15%	1.18
Cost per no.	£9.06

5

Tender settlement/adjudication

A previously discussed, the preparation of a bid or tender figure is a two-stage process; the calculation by the estimator of the true commercial cost to the contractor/sub-contractor followed by the adjudication or settlement process, or as NRM2 refers to it, the director's adjustment, where a number of factors are taken into account to arrive at the bid figure. These other factors may include some or all of the following:

- market conditions;
- type of contract;
- fixed price or fluctuations;
- non-measureable items; and
- perceived risk.

The impact of the above factors will be allowed for by adjusting risk allowances/ profit margins, etc.

NRM2 now gives the opportunity for a tendering contractor to show, in the summary to the bills of quantities, the amount of director's adjustment included in a tender as a percentage reduction or omission based on a commercial review of the project. Reaction to this has been mixed, with some contractors choosing not to include an amount on the grounds of commercial confidentiality while others take the view that a client would be persuaded to accept a tender if it can be demonstrated that money has been deducted from the original price i.e. the client's getting a bargain. A similar provision has been contained in the CESMM for a number of years, although the reason for the inclusion in these rules is rather more pragmatic.

MARKET CONDITIONS

Construction activity is cyclical; Figure 5.1 illustrates the dramatic swings in construction workload over the last two decades.

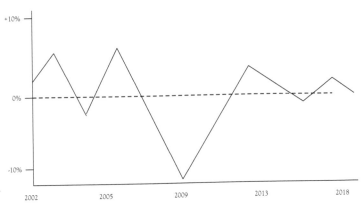

Figure 5.1 Construction output – percentage change 2002–2018

Source: BERR

According to various sources, these are some of factors influencing market conditions at the time of writing:

- Tender prices rose by 14.2% in 2017 but were expected to fall throughout 2018 (BCIS).
- Over the three months to September 2018, weekly earnings in construction averaged £746.00, 10% lower than the same period the year before (totaljobs.com).
- In 2018, average earnings throughout the whole economy rose by 2.8% over the same period mentioned above to £516 per week (Office for National Statistics).
- In the first quarter of 2018, approximately 2.96 million people were employed in construction (Office for National Statistics).
- Construction insolvencies soared 73% in the first quarter of 2018 compared with the final quarter (Q1) of 2017. There were 934 construction industry insolvencies in Q1 2018, up from 539 in Q4 2017 (Creditsafe).
- Labour shortages were at record levels in 2018, particularly bricklayers and carpenters (Federation of Master Builders).
- Material prices are expected to rise by 4–5% per annum in 2018 (Department of Business, Energy and Industrial Strategy).
- The lack of clarity in Brexit negotiations is, at the time of writing, resulting in a sharp fall in private commercial work.
- Some large general contractors were reporting a profits margin of less than 1% in 2017.

- The housing market remains buoyant.
- A total of six companies listed in the FTSE Construction and Materials category issued profit warnings for the first half of 2018 (EY's latest Profit Warnings Report).

When market condition are buoyant with plenty of work to be priced then contracting organisations will try to increase their profit margins in the confidence that if one bid is unsuccessful, like buses, another request to submit a bid will be along soon. Although contractors/sub-contractors are traditionally secretive about profit margins, it is thought that for general contracting the level of profit averages around 5% although house-building produces somewhat higher levels. Given the many risks involved in contracting, plus the fact that many UK contractors are one-product companies, without alternative income streams to help them boost profits in bad times, 5% is not that large. One of the questions to be addressed during the adjudication process is what level of profit should be added to the commercial cost? In times when work is scarce, the profit margin will almost certainly be reduced, as this is one of the only ways by which companies can easily establish true commercial advantage over their competitors.

The amount by which profit levels will be cut will depend on:

- the financial well-being of the company, which includes such items as the levels of cash flow needed to pay for wages, head-office costs, etc.;
- the relationship with the bank including overdraft levels and access to debt funding;
- the profit levels generated from other projects;
- the impact of loss-making projects; and
- the amount of existing long-term projects generating guaranteed cash flow.

During the last recession, RICS members reported that, when examining tenders, they were finding an alarming amount of unsustainable tender pricing. In some cases, reductions 25% below cost price were being received. This situation exists when contractors/sub-contractors reduce profit margins to negative levels, for example:

| True commercial cost of tender | £2,500,000.00 |
| Amount of submitted bid | £2,400,000.00 |

In the above situation, a contractor has elected to submit a bid that contains zero profit; in fact, the bid price is £100,000.00 less than the cost price of labour, materials, plant and overheads. This is done in the belief that the lowest price will win the job and when work starts there will be many opportunities to recoup costs and profit through variations/additions/extensions and claims for loss and expense. This approach is not new and can be used to generate and maintain cash flow, but only for a limited number of contracts and for a limited time. If unsustainable bids are submitted for all projects then this can cause serious financial problems.

TYPE OF CONTRACT

The 2018 NBS National Construction Contracts and Law Report indicates that in the UK the vast majority of construction contracts still use one of the standard forms of contract, albeit sometimes with client or consultant amendments. The JCT standard forms continue to dominate the construction contracts market with 70% of people surveyed indicating they use a JCT contract. NEC contracts continue to threaten the dominance of the JCT with a score of 39%. Bespoke contracts will generally carry greater risk and uncertainty.

As Table 5.1 illustrates, the UK construction industry likes to use standard forms of building contract. The reason is simply that clients and contractors are familiar with them and know how to refer to them in the case of a dispute between parties. Standard forms of contract have been described as a method of distributing risk and, as far as this is true, most experienced members of the industry seem happy with this state of affairs. If contracts were based on bespoke conditions then a risk analysis would have to be carried out for each new contract. That's not to say that some clients will not alter and amend the standard forms, usually in an attempt to transfer more risk to the contractor, a practice very much discourage by the publishers of standard forms and lawyers alike. Therefore, as far as the conditions of contract influencing the adjudication process is concerned, it's safe to say that this will only happen in the event of an unfamiliar form of contract being used or when the standard conditions have been amended by the client. The adjudication team should attempt to analyse and price the potential impact of non-standard or amended forms of contract conditions. Analysis and scrutiny of the contract is particularly important when bidding for work outside of the UK where different legal systems and conditions apply.

Table 5.1 Types of contract

Contract	Usage
JCT contracts	70%
NEC contracts	39%
Bespoke contracts	23%
RIBA contracts	14%
FIDIC contract	10%

Source: NBS National Construction Contracts and Law Report, 2018

FLUCTUATIONS

JCT (16) presents the following three possibilities for recovering increases in cost during the contract period:

- Option A – so-called firm price. The contractor is able to recover from the client increases in the cost of statutory payments and levies, such as National Insurance, etc.
- Option B allows the contractor to recover all increases in the prices of labour and materials and tax increases.
- Option C is the recovery of costs using the formula rather than traditional means.

A client may decide to ask for a contractor or sub-contractor to provide a firm price (Option A) for carrying out the works; that's to say the bidder has to absorb increases in cost due to rises in the cost of resources during the during of the project. In fact, firm price, in the case of JCT (16), is not fixed price as Option A allows the contractor to be reimbursed for the cost of increases in certain statutory charges, for example, statutory contributions, levies and taxation.

FIXED PRICE/FIXED-PRICE ADDITIONS

As well as the provisions for fixed price/fluctuations set out in the various standard forms of contract, there is also a fourth option. A client may ask a contractor or sub-contractor to submit a fixed-price addition to the submitted bid; that's to say a no-excuses additional lump sum to take account of all price risks during the contract period. The client can then choose to accept the lump-sum addition figure or decide that it is better to go with the standard forms standard fluctuation clauses. Obviously, the shorter the contract period and the lower the rate of inflation, the easier this task will be for the organisation submitting the bid. If a fixed-price bid is required, the estimator must calculate the amount of increases to labour, materials and plant for the contract period including any predicted statutory rises.

Labour

Of all the resources, labour is the easiest to predict as each year increases in the basic annual hourly rate are negotiated and the convention is that increases mirror the state of the market. For example, for April 2018 the all-in hourly rate for craftsmen 2018/19 based on the BATJIC agreement was £23.16 (see Chapter 3). The major economic forecasters' average forecast of retail price inflation is + 3.2% for 2018 and + 3% for 2019. The forecasts for 2018 range from + 2.4% to + 4% and for 2019 between + 2.5% and + 4.2%.

Assume a 24-month contract starting in September 2018:

All-in rate Sept. 2018–19	£23.16/hour
All-in rate Sept. 2019–20 + 3%	£23.86/hour
Average hourly rate	£23.51/hour

This rate should now be used in the re-calculation of bill/work-package rates priced according to when they will be carried out in the project programme to ascertain the probable increases in labour costs. If a single fixed-price addition figure is required by the client, then anticipated statutory increases should also be allowed for.

Materials

An accurate assessment of increases in the prices of materials during the contract period can be made by abstracting the quantities of all major materials and the time that they are going to be incorporated into the works. The estimator can then access a set of building-cost indices, such as those prepared by the BCIS. For bricks, for example, see Table 5.2. The impact of these indices can be assessed and applied to the quantities of bricks that are required for the job being priced.

Plant

Increases in the cost of plant can be assessed in the same way as materials. Even when using the provisions for recovering in the various standard forms and there is provision for contractors/sub-contractors to recover increases in cost, there is usually a shortfall in the costs recovered, the so-called non-recoverable costs. In order to calculate this figure, it is necessary to subtract the estimated increases in costs from the forecast amount recoverable under the contract, giving the non-recoverable element, which should be included in the bid price.

Table 5.2

Date	Index	Status	% variation: year	% variation: month
Q3 2016	280	Firm	– 16.7	5.7
Q4 2016	308	Firm	3.00	10.00
Q1 2017	303	Firm	0.30	– 1.6
Q2 2017	335	Firm	12.80	10.60
Q3 2017	311	Firm	11.10	– 7.20

Non-measureable items

A significant percentage of the items that form the bid price may not be included within measured items, bills of quantities/work packages, and there will, in addition, be less defined items that need to be considered and priced by the estimator. The greater the percentage of non-measured items included in the tender documents then the less influence the estimator can have over the bid as a whole.

In addition to the measured items in a bill of quantities or work package, there will be other cost centres that have to be considered during the estimating process. Typically, these items are:

- provisional sums;
- dayworks; and
- approximate quantities.

Provisional sums

Provisional sums are sums included in the bill of quantities/work package for any items of work that are anticipated, but for which no firm design has been developed, including any sums listed for any items of work that are to be executed by a statutory undertaker. This part of the tender documents, therefore, lists items of work that cannot be entirely foreseen or detailed accurately at the time tenders are invited (i.e. non-measurable items). Pre-determined sums of money are set against each item, calculated by the quantity surveyor/cost manager, to cover their cost and should, in value terms, be as close as possible to the eventual cost. For example:

Include the provisional sum of £10,000.00 for the supply
of garden equipment from a named supplier £10,000 00

As described in NRM2, there are two categories of provisional sum; defined and undefined.

DEFINED PROVISIONAL SUMS

A sum provided for work that is not completely designed but for which the following information shall be provided:

- the nature and construction of the work;
- a statement of how and where the work is fixed to the building and what other work is to be fixed thereto;
- a quantity (or quantities) that indicates the scope and extent of the work; and
- any specific limitations and the like identified.

As detailed above, a defined provisional sum provides the contractor/sub-contractor with a fair degree of detail and importantly, where provisional sums are included for defined work, the contractor/sub-contractor will be deemed to have made due allowance in the programming, planning and pricing preliminaries.

UNDEFINED PROVISIONAL SUM

- A sum provided for work that is not completely designed, but for which the information required for a defined provisional sum cannot be provided.
- Where provisional sums are given for undefined work, the contractor will be deemed not to have made any allowance in programming, planning and pricing preliminaries.

In the case of a defined/undefined provisional sum, as details are so sketchy they are exclusive of overheads and profit. Separate provision has to be made in the bid price for overheads and profit.

Prime-cost sum (first cost)

- A sum of money included in a *unit rate* to be expended on materials or goods from suppliers (e.g. ceramic wall tiles at £36.00/m^2 or door furniture at £75.00/door).
- It is a supply-only rate for materials or goods where the precise quality of those materials and goods are unknown.
- Prime-cost sums exclude all costs associated with fixing or installation, all ancillary and sundry materials and goods required for the fixing or installation of the materials or goods, sub-contractor's design fees, sub-contractor's preliminaries, sub-contractor's overheads and profit, main contractor's design fees, main contractor's preliminaries and main contractor's overheads and profit.

For example:

Include the prime-cost sum of £16.00/m^2
for wall coverings to committee rooms 69 m^2

The contractor/sub-contractor, as indicated above, will have to allow for the costs associated with preliminaries, overheads and profit associated with this item either as part of the bill/work package item or elsewhere in the tender documentation.

CONTRACTOR-DESIGNED WORK

Contractor-designed work, or as it is also referred to the contractor-designed portion (CDP), is work that requires the contractor, sometimes via a sub-contractor, to undertake the design of a portion of the works. This is usually identified in the bills of quantities or work packages. There are no hard and fast rules for quantifying CDP, it will very much depend on the nature of the works being carried out. It should, therefore, be borne in mind that CDP estimates shall be deemed to include all costs in connection with design, design management, and design and construction risks in connection with contractor-designed works, in addition to the following costs:

- labour and all costs in connection therewith;
- materials and goods together with all costs in connection therewith;
- assembling, installing, erecting, fixing or fitting materials or goods in position;
- plant and all costs in connection therewith;
- waste of goods or materials;
- all rough and fair cutting unless specially stated otherwise;
- establishment charges; and
- cost of compliance with all legislation in connection with the work measured including health and safety, disposal of waste and the like.

In addition, the contractor will be deemed to have made due allowance in their programming and planning for all design works in connection with contractor-designed works.

DAYWORK

Another mechanism for dealing with unforeseen or undefined work in the tender documentation is daywork. Daywork is a method of valuing work on the basis of time spent by the contractor's employees, the materials used and the plant employed. It is usually calculated using RICS/BEC *Definition of Prime Cost of Daywork Carried Out under a Building Contract, 3rd Edition* (June 2007); however, always check that this definition is the basis for the tender being prepared, as prime cost may be defined in other ways!

RICS/BEC definition of dayworks has two options for dealing with the prime cost of labour:

Option A: percentage addition

Option A is based upon the traditional method of pricing labour for daywork, and allows for a percentage addition to be made for incidental costs, overheads and profit, to the prime cost of labour applicable at the time the daywork is carried out. The composition of the total daywork charge will include:

- labour;
- materials and goods; and
- plant.

A schedule of dayworks is incorporated into the bill of quantities:

- Labour – list the various classifications of labour with estimates of the number of hours.
- Materials and plant – lump sums.
- Daywork rates and percentage addition are inserted by contractor/sub-contractor.

The prime cost of labour includes:

- wages and emoluments;
- additional WRA payments;
- general overhead costs; and
- project overhead costs.

To this is added a percentage addition to include:

- difference between cost of labour and prime cost;
- added cost of supervision/record keeping;
- disruption costs; and
- high cost of labour-only specialists.

Option B: all-inclusive rates

Option B is based on all-inclusive rates that include not only the prime cost of labour but also include an allowance for incidental costs, overheads and profit. The all-inclusive rates are deemed to be fixed for the period of the contract. However, where a fluctuating-price contract is used, rates shall be adjusted by a suitable index in accordance with the contract conditions. All-inclusive rates give the client price certainty in terms of the labour rate to be used in any daywork during the contract, but there is the potential that the rate will be higher, as the contractor is likely to build in a contingency to cover any unknown increases in labour rates that may occur during the contract period.

Prime-cost materials

- Cost from builder's merchant, including some discounts and delivery, unloading and storage.
- Typical percentage additions based on site (10%) and head-office overheads (5%) and profit (5%).

Prime-cost plant

The RICS/BEC definition includes a *Schedule of Basic Plant Charges* that is intended to apply solely for dayworks and not applied to jobbing or any other work, or work carried out during the rectification period. The plant charges include:

- all fuel;
- maintenance; and
- licences and insurances.

However, the following are excluded:

- drivers and attendants.

Percentage additions

Percentage additions will be included for:

- general overheads, incidental costs and profit
- the effect of inflation since publication of RICS/BEC schedule
- differences between rates in schedule and rates actually being paid.

There follows an example of allowing for daywork based on Option A of the RICS/BEC schedule.

Labour

The calculation of the prime cost of labour is different to the calculation of the all-in rate (see Chapter 3) as various incidental costs, overheads and profit are included in the percentage addition to the prime cost added by the contractor. The percentage addition should include all items that the contractor considers necessary to recover the true cost of carrying out works on a daywork basis. In accordance with RICS/BEC definition of dayworks, the prime cost of labour includes the following:

- guaranteed minimum wages and emoluments;
- additional emoluments in respect of the Working Rule Agreement; and
- overhead costs of employing operatives.

Plus a percentage addition to cover:

- incidental costs, overheads and profit.

Prime cost of labour (skilled) in accordance with RICS/BEC, *Definition of Prime Cost of Daywork Carried Out under a Building Contract, 3rd Edition,* Section 3.5:

Item	Quantity	Rate	£
(a) Standard weekly earnings	1,794 hours	£12.45	£22,335.30
(b) Extra payments for skill, etc.	Included	Included	£0.00
(c) Public holidays	63 hours	£12.45	£784.35
(d) Annual holidays	171 hours	£12.45	£2,128.95
(e) Death benefit	52 weeks	£7.50	£390.00
(f) Employer's levy and contributions			£88.37
(g) National Insurance contributions.	13.80% above threshold		£2,321.80
		÷ 1,794 hours	£28,048.77
		Prime-cost hourly rate	£15.64

To the prime cost of labour hourly rate (£15.64) must be added a percentage addition to cover the following:

(a) Head-office charges.
(b) Site staff including site supervision.
(c) The additional cost of overtime.
(d) Time lost due to inclement weather.
(e) The additional cost of bonuses and all other incentive payments in excess of any guaranteed minimum.
(f) Apprentices' study time.
(g) Subsistence, lodging and periodic allowances.
(h) Fares and travelling allowances.
(i) Sick pay or insurance in respect thereof.
(j) Third-party and employer's liability insurance.
(k) Liability in respect of redundancy payments to employees.
(l) Employer's National Insurance contributions not included in Section 3.5.
(m) Tool allowances.
(n) Use and maintenance of non-mechanical hand tools.
(o) Use of erected scaffolding, staging, trestles and the like.
(p) Use of tarpaulins, plastic sheeting and the like, all necessary protective clothing, artificial lighting, safety and welfare facilities, storage and the like that may be available on the site.
(q) Any variation to basic rates required by the contractor in cases where the building contract provides for the use of the specified *Schedule of Basic Plant Charges* (to the extent that no other provision is made for such variation – see Section 5).

(r) All other liabilities and obligations whatsoever not specifically referred to in this section nor chargeable under any other section.

(s) Any variation in welfare/pension payments from industry standard.

(t) Profit (including main contractor's profit as appropriate).

The all-in rate for skilled labour was previously calculated as £23.16 (see Chapter 3), therefore £7.52 or 48% (£23.16–£15.64) must be added to the cost to cover the costs of (a)–(r) listed above. In practice, a contractor/sub-contractor may add more to this percentage to cover the additional costs such as:

• additional supervision;
• working out of sequence of the original programme; and
• recording hours and record keeping.

Specialist trades such as M&E works have their own basis for calculating daywork rates, which will be discussed later. This will be allowed for in the bill of quantities/work package as described.

The contractor will be paid as defined below for the cost of works carried out as daywork in accordance with the building contract. For building works, the prime cost of daywork will be calculated in accordance with the latest *Definition of Prime Cost of Daywork (3rd Edition)* published by RICS and the Construction Confederation:

Building operatives	Provisional sum	£2,000.00
Add for incidental costs, overheads and profit …48…%		£920.00
Electrical operatives	Provisional sum	£1,000.00
Add for incidental costs, overheads and profit …………%		£
Heating and ventilating operatives	Provisional sum	£1,000.00
Add for incidental costs, overheads and profit …………%		£
Plumbing operatives	Provisional sum	£1,000.00
Add for incidental costs, overheads and profit …………%		£

Materials and goods

The prime cost of materials and goods obtained specifically for the daywork is calculated as follows:

• The invoice cost after deducting all trade discounts and any portion of cash discounts in excess of 5%, plus any appropriate handling and delivery charges.

- The prime cost of materials and goods supplied from the contractor's stock is based upon current market prices after deducting all trade discounts and any portion of cash discounts in excess of 5%, plus any appropriate handling charges.
- Any VAT is excluded for the purposes of calculation

As with labour, a percentage addition is added by the contractor. In the case of materials, this is an amount to cover for site and head-office overheads and profit. Typical amounts range from 10–20%.

Plant

Unless otherwise stated in the building contract, the prime cost of plant comprises the cost of the following:

- use or hire of mechanically operated plant and transport for the time employed/ engaged for the daywork;
- use of non-mechanical plant (excluding non-mechanical hand tools) for the time employed/engaged for the daywork;
- transport/delivery to and from site and erection and dismantling where applicable; and
- qualified professional operators (e.g. crane drivers) not employed by the contractor.

Where plant is hired, the prime cost of plant shall be the invoice cost after deducting all trade discounts and any portion of cash discount in excess of 5%. Where plant is not hired, the prime cost of plant shall be calculated in accordance with the latest edition of the RICS *Schedule of Basic Plant Charges for Use in Connection with Daywork under a Building Contract*.

The use of non-mechanical hand tools and of erected scaffolding, staging, trestles or the like is excluded (see Section 6).

Where hired or other plant is operated by the contractor's operatives, the operatives' time is to be included under Section 3 unless otherwise provided in the contract.

Any VAT is excluded, for the purposes of calculation.

The estimator will need to allow for various additions to the basic plant rates to cover the following:

- general overheads, incidental costs and profit;
- increases in cost since the publication of the *Schedule of Basic Plant Charges*; and
- the differences between rates given in the Schedule and the anticipated rates actually being paid for plant.

Daywork charges for specialist contractors and sub-contractors

As mentioned previously, in addition to formal definitions of daywork charges for general building works, the following definitions exist for so-called specialist trades:

- *Definition of Prime Cost of Daywork Carried Out under an Electrical Contract*, 3rd edition;
- *Definition of Prime Cost of Daywork Carried Out under a Heating, Ventilating, Air Conditioning, Refrigeration, Pipework and/or Domestic Engineering Contract*, 3rd edition;
- *Definition of Prime Cost of Daywork Carried Out under a Plumbing Contract*, 1st edition; and
- *Definition of Prime Cost of Building Works of a Jobbing or Maintenance Character*, 2nd edition.

As with the calculation of the prime cost for building work, the calculation of the prime cost for electrical work is not as comprehensive as the calculation of the all-in hourly rate.

The following is an example of the prime cost of labour for daywork in accordance with the JIB working rules for an approved electrician (outside London) and an electrician working in London:

	Rate	Rate	Approved electrician (outside London)	Rate	Electrician (London)
Basic salary	1,725 hours*	£21.16	£36.501.00	£20.73	£35.759.25
Public holidays	60 hours	£21.16	£1,269.60	£20.73	£1,243.80
Sub total			£37,770.60		£37,003.05
National Insurance**			£3,874.63		£3,772.27
Holidays with pay	165 hours	£21.16	£3,491.40	£20.73	£2,420.45
Annual labour cost	÷ 1,725 hours		£45,136.63		£44,195.77
	Hourly base rate		£26.17		£25.62

* Annual working hours calculated as follows:

52 weeks @ 37.5 hours	1,950
Less:	
Annual holiday	165
Public holiday	60
	1,725

** Threshold and percentage for National Insurance vary from time to time – see HMRC website.

Fluctuations

During the life of a building contract, the cost of labour and materials will increase and the longer the contract, the more significant this increase will be. The JCT (16) Standard Form of Building Contract has three options for dealing with increase costs or fluctuations:

- Option A: this is sometimes referred to as the firm-price option, although, in practice, increases in certain statutory changes such as government levies and taxes are refunded to the contractor.
- Option B: this option reimburses increases or decreases in the cost of labour and materials and is referred to as full fluctuations.
- Option C: this is the formula method that uses average-price indices to calculate increases in costs.

Option A – to strike out the fluctuations clauses completely and place the risk for covering increased costs entirely with the contractor – is a very popular option as it gives the client more certainty, particularly in times of high inflation.

If Option B is used then, at the time of the preparation of the tender, a list of basic materials should be compiled by the contractor. This list is referred to at the preparation of interim valuations during the contract period and compared with the actual cost being paid for materials, with any differences being paid to the contractor. This approach invariably results in more work for the contractor and sub-contractors when preparing interim valuations.

Option C – the final option – reimburses the contractor, not on the basis of the actual costs of materials and labour but on the changes in cost indices published each month by the fluctuations by Her Majesty's Stationery Office. A non-adjustable element, usually 10%, is stated in the contract.

Another approach is that the client asks the contractor to submit alternative prices for fixed and fluctuating terms; in this case, the contractor must calculate a fixed price addition, that is, the anticipated costs of increased costs for the duration of the contract.

Notwithstanding the above, there is usually a shortfall in the recovery of increased costs due to inaccuracies in:

- forecast increases in the cost of resources; and
- forecast amounts to be recovered.

Approximate bills of quantities

A contractor or sub-contractor will sometimes be asked to submit a tender based on an approximate bill of quantities/work package. It may be that the whole of the bill or work package is approximate, or just sections, for example, the substructure.

Approximate bills of quantities are used when there is insufficient detail to prepare firm bill of quantities or where it is decided by the employer that the time or cost of a firm bill of quantities is not warranted. Such contracts do not provide a lump-sum price, but rather tender-price totals, since the quantities are subject to re-measurement on completion by the quantity surveyor. These contracts are usually subject to greater variation than lump-sum contracts and therefore should only be used where time is a limiting factor or where there is great uncertainty in respect of certain elements, such as major excavation and earthworks. The initial resource cost of an approximate bill of quantities is likely to be lower than for a firm bill of quantities, but the need for re-measurement invariably results in an overall higher resource cost and this should be borne in mind by the estimator during the pricing process.

Overheads

As briefly discussed in Chapter 1, the preparation of a bid by a contractor or sub-contractor is a two-stage process. Previously in Chapters 3 and 4, calculating the true commercial cost of a project was explained; however, in addition to these costs the following items must be added to complete the bid:

- general overheads; and
- profit.

Overheads fall into two main categories:

- Project-specific overheads: these items are usually included within the Preliminaries section of the bill of quantities or work package and can include such items as site accommodation, site security, etc. These are dealt with in Chapter 4 – Unit-rate pricing.
- General overheads: these are items such as head-office overheads and salaries, etc. A percentage must be added to each job in order to contribute to general overheads to keep the contractor's organisation in business. The percentage required will vary from organisation to organisation and can be calculated on a previous year's turnover as follows:

  o last year's turnover: £20,000,000;
  o fixed costs: £1,600,000.

$$\frac{1,600,000}{20,000,000} \times 100 = 8\%$$

In this example, 8% should be added to the true commercial cost of bill items in order to recover the general overheads of the contractor/sub-contractor.

RISK

> There are known knowns; there are things we know we know. We also know
> there are known unknowns; that is to say we know there are some things we
> do not know. But there are also unknown unknowns – the ones we don't know
> we don't know.

(Former United States Secretary of Defense,
Donald Rumsfeld, 2002)

A widely accepted definition of risk is *an uncertain event or set of events that, should
it occur, will have an effect on the achievement of the project objectives*. In the context
of estimating and tendering, the achievement of the project objectives means not
realising the projected profit margins or completing the project within the stipu-
lated contract period. It is therefore essential that an effective risk-management
model is baked into the estimating/bid process. The construction industry is lit-
tered with contractors and sub-contractors who, for whatever reason, decided to
ignore the impact of risk and subsequently went out of business. To end this chapter on risk, and as if to emphasise the importance of effective risk management
during the estimating process, the following example should prove to be a salutary
case study. In the last half of the twentieth century John Laing Construction was
a household name in UK construction with many prestigious projects to its name.
However, in apparent confidence that it could handle whatever risks came its way,
the company seemed to put aside risk management and submitted bids for – and
won – two large projects:

- the Millennium Stadium in Cardiff; and
- the National Physical laboratory.

However, unfortunately for Laing, both of these projects were high-risk D&B projects, and the risks were not detected during the bid process. Subsequently, Laing
incurred massive losses on both contracts and in 2000 John Laing Construction was
sold for £1 to its main concrete-work sub-contractor.

Risks may be classified as:

- commercial risk; or
- technical risk.

Risk can be influenced by both internal and external factors and is broadly
divided into;

- client risk; or
- contractor risk.

Consequential risk

There are often interrelationships between risks, referred to as consequential risks, which increases the complexity in assessing risk. It is not uncommon for one risk to trigger or increase the impact and/or likelihood of another. A decision tree is a technique for determining the overall risk associated with a series of related risks.

The questions that must be addressed are as follows:

- identifying – what are the risks?
- analysing – what their impact will be?
- planning – what is the likelihood of the risks occurring?

Contractors and sub-contractors face two problems when pricing and submitting a tender, namely:

- Will the tender be sufficiently competitive to be accepted by the client?
- If selected, will the final outcome in terms of levels of profit be close to the predicted levels included during the adjudication stage?

As previously discussed, by its very nature, construction projects are risky undertakings, with contracts let as the result of competitive tendering being perhaps the riskiest of all. In Chapter 1 various procurement strategies were discussed as were the ways in which risk is distributed for various procurement routes.

Despite the introduction of systems such as Building Information Management (BIM), no matter how well planned and resourced, all construction projects contain an element of risk due to exposure to such externalities as:

- inclement weather;
- strikes and industrial disputes; and
- difficulties in obtaining critical materials.

Prior to submitting a bid, a contractor or sub-contractor must evaluate all risks and uncertainties, and incorporate an appropriate allowance in order to safeguard profit margins and mark-up.

The probabilities of some events or risks occurring can, in principle, be calculated, but in other cases the chances of an event occurring is inherently unknowable. This distinction is critical to practical approaches to quantifying risk, but it is not so relevant to the question of which risks should be considered as candidates for transfer. In practice, risk may be assessed using quantitative techniques but commercial judgement and political considerations will ultimately always have a part. During the bidding stage, risk may be managed in a number of ways ranging from complete transparency where risk is identified and apportioned prior to the pricing process to a situation where each player is expected to identify and manage risk in isolation. Any model attempting to analyse the impact of risk should recognise that

- risk relates to a specific event or set of events within a defined time frame;
- it is an estimation of the probability of the event(s) occurring; and
- the consequences of the risk should be capable of measurement.

Contractors may take the Rumsfeld approach when preparing a bid and choose to classify risk as:

- quantifiable risk; and
- unquantifiable risk – uncertainty.

Uncertainty arises when the probability of the occurrence or non-occurrence of an event is indeterminable and not assessable. While there is no clear demarcation between risk and uncertainty, some events may be transferred from being thought of as uncertainty to risk by collection of more comprehensive information. In understanding risk and uncertainty, it is therefore important to identify:

- whether it is possible to obtain further information;
- what's the cost of obtaining the information; and
- having discovered the additional information, whether it will be possible to analyse it and use it productively.

Risk evaluation

Once potential risks have been identified, some mechanism is required to quantify and evaluate them. Some very sophisticated risk-analysis and risk-management techniques are available that are widely used in other sectors. There is, however, some evidence that the techniques used in construction are somewhat more basic. A risk register is perhaps the most open and transparent approach to risk management. Some of the items that can be included in a register are illustrated in Tables 5.3 and 5.4.

A risk register will generally:

- identify risks that are capable of being identified;
- assess the probability of the risk occurring;
- develop a range of possible outcomes (worst case, medium case and best case) for each risk;
- value the outcome and the timing of each risk;
- assign probabilities to each outcome;
- calculate the expected value of each risk as the weighted average value of probability of the risk occurring, the outcome of the values and their probabilities; and
- finally, value each risk.

Risk registers are perhaps most critical in large, complex schemes such as a private finance initiative project where the risks have to be identified and priced at tender stage.

Other risk-identification techniques include:

- brainstorming;
- cause-and-effect diagrams;
- SWOT analysis;
- post-project reviews/lessons learned;
- questionnaires;
- project documentation review; and
- sensitivity analysis.

Qualitative risk analysis

The purpose of qualitative analysis is to prioritise the risks in terms of importance, without quantifying (costing) them. An assessment is made of the likelihood that the risk will occur and the magnitude of its potential impact. The qualitative severity rating is arrived at by multiplying the likelihood of occurrence by the qualitative impact. Likelihoods and impacts can be categorised as follows:

Likelihood	Probability
5 Very high – almost certain to occur	75–99
4 High – more likely to occur than not	50–74
3 Medium – fairly likely to happen	25–49
2 Low – but not impossible	5–24
1 Very low – unlikely	0–4

Impact on project costs	
5 Very high – critical impact on cost	2.00%
4 High – major impact on cost	1.50%
3 Medium – reduces feasibility	1.00%
2 Low – minor loss	0.50%
1 Very low – minimal loss	0.25%

Therefore, if a medium likelihood (3) is multiplied by a high impact on cost (4), the result is a total rating of 12.

Had John Laing construction carried out such an exercise on the design and building of the complex sliding roof of the Millennium Stadium, then perhaps they would still be around today.

Table 5.3 Design/construction project sample risk list

Construction risks
Unidentified utility impacts
Unexpected archaeological findings
Changes during construction not in contract
Unidentified hazardous waste
Site is unsafe for workers
Delays due to traffic management and road closures

Design risks
Incomplete quantity estimates
Insufficient design analysis
Complex hydraulic features
Surveys incomplete
Inaccurate assumptions during the planning phase

Environmental risks
Unanticipated noise impacts
Unanticipated contamination
Unanticipated barriers to wildlife
Unforeseen air-quality issues

External risks
Project not fully funded
Politically driven accelerated schedule
Public-agency actions cause unexpected delays
Public objections
Inflation and other market forces

Organisational risks
Resource conflicts with other projects
Inexperienced staff assigned to project
Lack of specialised staff
Approval and decision processes cause delays
Priorities change on existing programmes

Project-management risks
Inadequate project scoping and scope creep
Consultant and contractor delays
Estimating and/or scheduling errors
Lack of co-ordination and communication
Unforeseen agreements required

Right-of-way (ROW) risks
Unanticipated escalation in ROW values
Additional ROW may be needed
Acquisition of ROW may take longer than anticipated
Discovery of hazardous waste during the ROW phase

Table 5.4 Generic project sample risk list

Technical, quality or performance risks

Examples include reliance on unproven or complex technology, unrealistic performance goals, long-term performance, process roadblocks, new emerging initiatives, increases in complexity, etc.

External risks

Examples include a shifting regulatory environment, labour issues, changing customer priorities, government-agency risks, and weather. Also to be considered are consultant and vendor contract risks, contract type and contractor responsibilities.

Organisational risks

Examples include lack of prioritisation of projects, inadequacy or interruption of funding, inexperienced and poorly developed and trained workforce, and resource conflicts with other projects in the organisation.

Project-management risks

Examples include poor allocation of time and resources, inadequate quality of the project plan, lack of project-manager delegated authority, and lack of project-management disciplines.

Pricing risk

The nature of the risks as well as their allocation will vary from sector to sector and to some extent from project to project within the same sector. The construction of a risk matrix usually comprises the following steps:

- identify all risks involved in a project;
- assess the impact of these risks;
- assess the likelihood of the identified risks occurring; and
- calculate the financial impact when risks occur – this should be done over a variety of possible outcomes.

Risk workshops should be held comprising key stakeholders to identify the various risks. The usual approach is to prepare a risk register. The financial impact of the risks, some of which will be retained by the client and some of which will transferred to the client, will then be dealt with as follows:

- identify the risk that are capable of being quantified;
- assess the probability of the risk occurring;
- develop a range of possible outcomes (worst case, medium case and best case) for each risk;
- price the outcome and the timing of each risk;
- assign probabilities to each outcome; and
- calculate the expected value of each risk as the weighted average value of probability of the risk occurring, the outcome values and their probabilities.

The relationship between risk, value for money and affordability is one of the most complex areas due to the inherent risks in construction related projects. The longer that money is exposed to potential risks, the greater the chance that uncertainties will affect the investment. Rather like over taking a large slow-moving vehicle in a car, the quicker the manoeuvre can be completed, the less exposure to risk.

NRM2 risk register

NRM2 includes a provision for clients to include a schedule of risks (Table 5.5) in the bills of quantities. The schedule gives the client the opportunity to ask the contractor to price risks that the client wishes the contractor to take responsibility for. The risks are identified by the client and fully described so that it is transparent what risk the contractor is required to manage, and what the extent of services and/or works the employer is paying for. In the event that the risk does not materialise the contractor retains the sums that have been entered.

Operational risks

Some procurement strategies task the contractor with not only determining the capital cost of the project but also estimating long-term costs.

Following the identification and valuation of risk, the next stage in the evaluation process is to perform a sensitivity analysis, the purpose of which is to identify and value factors or events over the life of the contract. The technique measures the impact on project outcomes of changing one or more key input values about which there is uncertainty. For example, a pessimistic, expected and optimistic value might be chosen for an uncertain variable such as the demand in 15 years' time for the service, or the cost of maintenance. Then an analysis could be performed to see how the outcomes changes as each of the three chosen values is considered in turn, with other things held the same. Sensitivity analysis measures the economic impact resulting from alternative values of uncertain variables that

Table 5.5 Schedule of risks

Cost centre	Risk description	£/p
R001		
R002		
R003		
Etc.		
Total Risk Allowance, exclusive of VAT (carried to main summary)		£

affect the economics of the project. The results can be presented in the form of text, tables or graphs. This is a useful way of asking the 'what if' question. Not only is sensitivity analysis used in the preparation of the business case, it will also be used by private-sector consortia to calculate the possible impact of risks that they are required to manage.

Consider the choice of heating systems and controls for a school project. A decision must be taken on whether to install an expensive programmable system to control heating and ventilation. The control system will reduce consumption of energy by turning off equipment not needed when the building is unoccupied. The cost of the system will be justifiable if the net present value of future savings is greater than the cost of the new equipment. It is a relatively straightforward costs-in-use calculation to calculate the costs of the control system (purchase and installation) and the consumption of energy; however, the amount of savings that may accrue are not so certain as they are particularly susceptible to the price that will have to be paid for energy over the life cycle of the contract. Therefore, to test the sensitivity of the savings, three values of energy price changes are considered: low, moderate and high price increases. These can then be used to calculate possible outcomes, that's to say cost of new controls minus energy costs:

1. Low energy-price rises   (£20,000)
2. Medium energy-price rises   £20,000
3. High energy-price rises   £50,000

In the first case, assuming energy prices increases are minimal, the new equipment would actually cost £20,000 more than the projected savings; however, in the case of medium or high price increases considerable savings could be made. Note that this analysis contains no indication of the likelihood of increases occurring; however, if even moderate increases were predicted then the controls should be installed.

There are several good reasons to use sensitivity analysis during the preparation of operational costs. First, it shows the significance of project variables on the profitability of the project and well as identifying critical inputs in order to facilitate choosing where to spend extra resources in data estimates and in improving data estimates. Second, the technique is useful in preparing for the 'what if' question and accessing the robustness of the business case. It does not require the application of probabilistic techniques and, finally, it can be used when there are little data and time available. The major disadvantage is, however, that there is no probabilistic measure of risk exposure; although it may be fairly reasonable to expect one of several outcomes, the analysis contains no explicit measure of their respective likelihoods. There are, however, a number of proprietary software packages available that claim to combine probability functions within sensitivity analysis.

NRM and the approach to risk

NRM1 and 2 have both attempted to make the way in which the construction industry deals with risk more transparent to all parties. It is recommended that the amounts to cover risks are quantified and entered into estimates, cost plans and bills of quantities/work packages.

Submission of the completed tender

On completion of the preparation of the bid, the price for carrying out the works is submitted to the client/architect/quantity surveyor depending on the requirements of the tender documents.

6

The supply chain

THE CHANGING NATURE OF THE CONSTRUCTION INDUSTRY

Forty or so years ago, the structure of the UK construction industry was very different to the modern day, with a majority of general contractors employing a range of tradesmen directly, with generally only specialist works such as M&E installations being carried out by sub-contractors. However, many modern main contractors are now simply managing contractors; that's to say, they manage sub-contract works carried out by a range of organisations that are not part of their organisations. The rise of the importance of the sub-contractor can be attributed to the following factors.

THE SUPPLY CHAIN

The construction supply chain may be considered as a network of organisations involved in the different processes and activities that produce the materials, components and services that come together to design, procure and deliver a building.

The supply chain that is traditionally assembled to carry out a construction project is known as a temporary multi-organisation. In many cases, the people involved in the design and construction processes differ from one project to another, with the construction team being formed and reformed for each new project. It must very quickly go through all the usual 'team-building' processes if it is to operate as an efficient working organisation.

Figure 6.1 illustrates part of a typical construction supply chain, although in reality many more sub-contractors could be involved. The problems for process control and improvement that the traditional supply-chain approach produces are related to:

- the various organisations coming together on a specific project, at the end of which they disband to form new supply chains;
- communicating data, knowledge and design solutions across the organisations that make up the supply chain;

- stimulating and accumulating improvement in processes that cross the organisational borders;
- achieving goals and objectives across the supply chain; and
- stimulating and accumulating improvement inside an organisation that only exists for the duration of a project.

Where construction is concerned, the supply chain for the delivery of a product clearly includes:

- the main contractor;
- other general or specialist contractors that the main contractor may employ to assist in carrying out the works;
- suppliers of materials and products to be incorporated in the works; and
- suppliers of professional services such as architects, consulting engineers, etc.

Where manufacturing industries are concerned, the supply chain stops at the point of delivery to a customer who will have a choice between various alternative but similar products. By comparison, in construction it is usually the client who is the ultimate consumer of the produce and has a large influence on the design and form of the finished product. Therefore, in construction there is a growing body of opinion that regards the client as an integral part of the supply chain. Interestingly, the long history of mistrust between main contractors and the sub-contractors in the construction industry has left many small sub-contractors looking to the client to defend their interests within the supply chain.

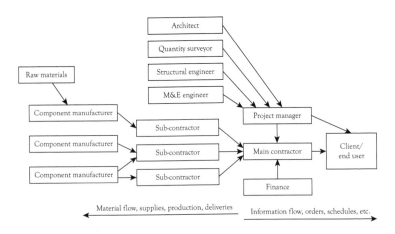

Figure 6.1 The construction supply chain

SUB-CONTRACTORS

Much of the work attributed to a general contractor is carried out by sub-contractors. Sub-contractors can be individuals or substantial firms who agree a contract with the main contractor to complete a section of a project, for example ground-works and masonry. During the recent past, the percentage of work being carried out by sub-contractors has increased considerably. For the main contractor, the advantages are:

- it reduces the main contractor's liability to retain, on a full-time basis, all the specialists necessary for day-to-day operations;
- it transfers risk to the supply chain;
- sub-contractors are used as and when required, thereby reducing overheads;
- it avoids having to pay operatives when work is scarce or having to pay redundancy; and
- it introduces the potential for innovation.

The disadvantages to employing sub-contractors are:

- programming: some types of sub-contractors may, on occasions, be difficult to engage;
- non-availability of the best sub-contractors;
- less control of health and safety issues;
- control and co-ordination of sub-contractors;
- quality of work: and
- potential for clashes of culture and philosophy.

It also needs to be taken into consideration that standard forms of contract vary in the amount of information they are required to provide about sub-contractors, and the degree of authority vested in those responsible for giving permission for work to be sub-contracted.

There is a number of different types of sub-contractor as follows:

- domestic sub-contractors;
- named sub-contractors; and
- nominated sub-contractors (although these are increasingly less common).

Domestic sub-contractors

Domestic sub-contractors are employed directly by the main contractor; it is a private arrangement between the two parties. Domestic sub-contractors do not have a contract with the client, but work on site as if they are the main contractor's personnel and they are co-ordinated by the main contractor's site-management team.

Domestic sub-contractors have a responsibility to deliver work that complies with the approval of the client and architect/contract administrator and are paid by the main contractor from monies received from the client in interim valuations.

The terms of payment, discounts, etc. are negotiated between the domestic sub-contractor and the main contractor. Although the various forms of contract vary slightly, as long as the main contractor informs the architect/contract administrator that certain parts of the works are to be carried out by domestic sub-contractors, consent cannot be reasonably withheld.

Domestic sub-contractors are commonly used as follows:

- to carry out complete sections of the work, including supplying all materials; and
- for labour and plant.

Labour-only sub-contractors, as the title suggests, supply only labour to the main contractor. Fears have been expressed over the use of this sort of labour, particularly over:

- lack of training and skills;
- accountability; and
- payment of taxation and other statutory obligations.

It may be helpful to realise the contractual relationship in the case of the above two types of sub-contracting is between the main contractor and the sub-contractor only. The main contractor retains the responsibility for ensuring domestic sub-contractors comply with all relevant statutory legislation and is answerable to the architect/contract administrator for the works done and materials supplied.

Named sub-contractors

Named sub-contractors were first introduced in the 1998 JCT Intermediate Form of Contract in an attempt to reduce the perceived complexity of the nomination process. In 2005, named sub-contractors replaced nominated sub-contractors in JCT (05).

Named sub-contractors are often used in public-sector projects and for projects based on the form of contract where there is no provision for nomination. If a named sub-contractor has not been selected at the time that the tender documents are ready for despatch then a provisional sum is included for provision of the works to be carried out by the named sub-contractor. The main features of named sub-contracting are as follows:

- The tender documents include the names of potential named sub-contractors. The main contractor has the opportunity to reasonably object to any firm on the list.

- The main contractor leads the tender process for each named sub-contractor package by assembling the tender documents, issuing and receiving tenders, and selecting a named contractor.
- After appointment, the named sub-contractor is, for all intents and purposes, a domestic sub-contractor. The main contractor is only paid the rates in the accepted sub-contract tender.

Because the contractual relationship between a named and a domestic sub-contractor is similar, the named sub-contractor is allowed for by the inclusion of a provisional sum.

Nominated sub-contractors

A small number of standard forms of contract include the provision for appointing nominated sub-contractors in cases where the architect/contract administrator or client wishes to restrict and control certain aspects of the project works. It may also be used in cases where, at the tender stage, parts of the project have not been fully detailed and therefore the use of nominated sub-contractors allows the job to go to tender, with the nominated works being dealt with at a later date. The principal differences between domestic and nominated sub-contractors are:

- the tender process is organised and run by the architect/contract administrator who invites suitable sub-contractors to submit a tender; and
- the architect/contract administrator selects the preferred tender and then instructs the main contractor to enter into a contract with the sub-contractor.

The client has no direct contractual relationship with a nominated sub-contractor as the sub-contract itself is still placed by the contractor. However, a client who wishes to use a nominated sub-contractor may obtain a collateral warranty from the sub-contractor so that there is a line of recourse against the sub-contractor where the client retains responsibility for delay or defects caused directly by the sub-contractor.

SELECTING QUOTATIONS

Sub-contractors

As such a high percentage of work is carried out by sub-contractors and suppliers, it is obviously vitally important that contractors have robust systems in place for selecting and appointing sub-contractors and suppliers, as the quality of the final building will depend on their expertise and professionalism. Most contractors have lists of sub-contractors and suppliers with whom they have worked previously and have confidence in their ability to turn out a good job. However, there may be occasions when, say, due to the uniqueness of the design or perhaps working in an unfamiliar geographical location, new and untried sub-contractors will have to be selected and appointed.

The quality and the accuracy of sub-contractor's and supplier's prices/quotations will very much depend on the quality of the information available for them to prepare quotations. The process of obtaining sub-contractors quotations is as follows:

- A standard enquiry letter (see Figure 6.2) should be drawn up and sent to selected sub-contractors.
- It is important that the information sent to sub-contractors/suppliers is as accurate as possible,
- The usual approach is to split the bill of quantities into work/trade packages, the format of NRM2 aids this process as it is arranged in work-package format.
- In addition to the above, the sub-contractor should be supplied with relevant sections from the specification together with drawings.
- It is important in the covering letter to make clear what services and facilities will be provided by the main contractor – there are often disputes between main contractors and sub-contractors as to what attendances are to made available to sub-contractors; these typically may include:

  o unloading and storage;
  o power supply;
  o removal and clearing from site of packaging, protection, etc.;
  o temporary accommodation; and
  o setting out from main contractor's base lines.

  The nature of main-contractor attendances will vary from sub-contractor to sub-contractor and should not be assumed to be standard. For example, when checking a sub-contractor's quotation for curtain walling, the sub-contractor specifically excludes the cost of removing and disposing of hundreds of square metres of protective tape from the face of the curtain walling, once installed. In this case, the sub-contractor should be asked to either include for the disposal of protection in the price or the main contractor must allow for the item in the preliminaries.
- Queries raised by sub-contractors during the tender period are dealt with in the same manner as with main contractor queries on the bills of quantities.
- Once a sub-contractor has been selected, the results should be communicated to the firms that submitted a quotation.

Suppliers

In general, the selection of suppliers of materials will be less rigorous than for sub-contractors; this is because suppliers will be less concerned with the type of the project or any restrictions in working conditions.

To;

Sub-contractor   Date

Dear Sirs,

New School, Dorking, Surrey.

We refer to our recent telephone conversation in connection with the above project
and invite you to submit a tender for the supply and installation of suspended ceilings
in accordance with the following details.Please find attached the following;

        Drawings nos. 123D to 128F inclusive
        Preliminary clauses
        Specification of suspended ceilings
        Tagged pages from bills of quantities nos. 6/01 – 6/15
        Daywork details
        Health & safety plan

The completed tender should be returned by – *time and date inserted.*

The contract will be JCT DOM/1 (2011) completed as follows:

        Payments – monthly
        Discount to main contractor – 2½%
        Fluctuations – firm price
        Liquidated damages – £1500.00 per week
        Retention – 3%
        Rectification period – 6 months

The following items will be provided free of charge;

Water, lighting and electrical services
Welfare facilities

Sub-contractors will be required to provide the following:
Unloading, storing and distribution around the site
Clearing away and removing all rubbish and packaging
Temporary accommodation and telephones

To arrange a site visit please contact.................................... at ....

Yours faithfully,

ABC Builders plc

Figure 6.2 Sub-contractor enquiry letter

MANAGING THE SUPPLY CHAIN

What is a supply chain?

Supply-chain management has its origins in the *keirstsu*, the *kanban* and the ship
yards of Japan in the 1950s, being further refined by Taiichi Ohno of Toyota some
20 years later. In a construction context, the contracts and partnerships that create the temporary organisation that delivers a project are illustrated in Figure 6.1.
Before establishing a supply chain or supply-chain network, it is crucial to understand fully the concepts behind, and the possible components of, a complete and
integrated supply chain. The term supply chain has become used to describe the

sequence of processes and activities involved in the complete manufacturing and distribution cycle – this could include everything from product design through materials, and component ordering through manufacturing and assembly until the finished product is the hands of the final owner. Of course, the nature of the supply chain varies from industry to industry. Members of the supply chain can be referred to as upstream and downstream supply-chain members as Figure 6.3. Supply-chain management, which has been practised widely for many years in the manufacturing sector, therefore refers to how any particular manufacturer involved in a supply chain, manages its relationship both up- and downstream with suppliers to deliver cheaper, faster and better end products. In addition, good management means creating a safe commercial environment, in order that suppliers can share pricing and cost data with other supply-chain team members.

The more efficient or lean the supply chain, the more value is added to the finished product. As if to emphasise the value point, some managers substitute the word value for supply to create the value chain. In a construction context, supply-chain management involves looking beyond the building itself and into the processes, components and materials that make up the building. Supply-chain management can bring benefits to all involved when applied to the total process, which starts with a detailed definition of the client's business needs. This can be provided through the use of value management and ends with the delivery of a building that provides the environment in which those business needs can be carried out with maximum efficiency and minimum maintenance and operating costs. In the traditional methods of procurement, the supply chain does not understand the underlying costs, hence suppliers are selected by cost and then squeezed to reduce price and whittle away profit margins. It tends to work best in industries in which: jobs or business opportunities are episodic and somewhat unpredictable rather than continuous; the required capability or skill mix varies from job to job; and, finally, where the costs of retaining a full spectrum of skills cannot be justified. For most industries, however, building long-term relationships based on trust and a high level of integration yields greater benefits. Construction performance can often be maximised through the nurturing of long-term relationships,

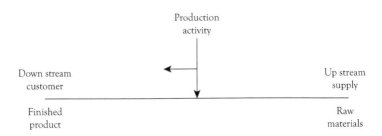

Figure 6.3 Supply-chain activity

even if the skill set is only used on a contract-by-contract basis. How a prime contractor selects contractors and suppliers is very much down to the individual organisation and procurement strategy. Traditionally, the approach to managing sub-contractors was:

- bids based on designs to which suppliers have no input, no buildability;
- low bids always won;
- unsustainable – costs recovered by other means;
- margins low, so no money to invest in development; and
- suppliers distant from final customer so took limited interest in quality.

The immediate implications of supply-chain management are:

- key suppliers are chosen on criteria, rather than job by job by competitive quotes;
- key suppliers are appointed on a long-term basis and proactively managed; and
- all suppliers are expected to make sufficient profits to reinvest.

Not all of the techniques are new; many practising quantity surveyors would agree that the strength of the profession is expertise in measurement, and in supply-chain management, there is a lot to measure. For example:

- productivity – for benchmarking purposes;
- value – demonstrating added value;
- out-turn performance – not the starting point;
- supply-chain development – are suppliers improving as expected?; and
- ultimate customer satisfaction – customers at supermarket, passengers at airport terminal, etc.

The traditional construction project supply chain can be described as a series of sequential operations by groups of people or organisations.

Supply chains are unique, but it is possible to classify them generally by their stability or uncertainty on both the supply side and the demand side. On the supply side, low uncertainty refers to stable processes, while high uncertainty refers to processes that are rapidly changing or highly volatile. On the demand side, low uncertainty would relate to functional products in a mature phase of the production life cycle, while high uncertainty relates to innovative products. Once the chain has been categorised, the most appropriate tools for improvement can be selected.

The construction supply chain is the network of organisations involved in the different processes and activities that produce the materials, components and services that come together to design, procure and deliver a building. Traditionally, it is characterised by lack of management, little understanding between tiers of other tiers' functions or processes, lack of communication and a series of sequential operations by groups of people who have no concern about the other groups or the client. However, supply-chain management takes a different approach that includes the following:

- Prices are developed and agreed, subject to an agreed maximum price with overheads, and profit is ring fenced. All parties collaborate to drive down cost and enhance value with, for example, the use of an incentive scheme.
- With cost determined and profit ring fenced, waste can now be attacked to bring down price and add value with an emphasis on continuous improvement.
- As suppliers account for 70–80% of building costs, they should be selected on their capability to deliver excellent work at competitive cost.
- Suppliers should be able to contribute new ideas, products and processes, and build alliances outside of project, and be managed so that waste and inefficiency can be continuously identified and driven out.

The philosophy of integrated supply-chain management is based upon defining and delivering client value through established supplier links that are constantly reviewing their operation in order to improve efficiency. There are now-growing pressures to introduce these production philosophies into construction and it is quantity surveyors with their traditional skills of cost advice and project management who can be at the forefront of this new approach. For example, the philosophy of lean thinking, which is based on the concept of the elimination of waste from the production cycle, is of particular interest in the drive to deliver better value. In order to utilise the lean-thinking philosophy, the first hurdle that must be crossed is the idea that construction is a manufacturing industry that can only operate efficiently by means of a managed and integrated supply chain. At present, the majority of clients are required to procure the design of a new building separately from the construction; however, as the subsequent delivery often involves a process where sometimes as much as 90% of the total cost of the completed building is delivered by the supply-chain members, there would appear to be close comparisons with, say, the production of a motor car or an airplane.

The basics of supply-chain management can be said to be:

- determine which are the strategic suppliers, and concentrate on these key players as the partners who will maximise added value;
- work with these key players to improve their contribution to added value; and
- designate these key suppliers as the 'first tier' on the supply chain and delegate to them the responsibility for the management of their own suppliers, the 'second tier' and beyond.

To give this a construction context, the responsibility for the design and execution of, say, mechanical installations could be given to a 'first tier' engineering specialist.

This specialist would in turn work with its 'second tier' suppliers as well as with the design team to produce the finished installation. Timing is crucial as 'first tier' partners must be able to proceed confident that all other matters regarding the interface of the mechanical and engineering installation with the rest of the project have been resolved and that this element can proceed independently. At least one food retail organisation using supply-chain management for the construction of its stores still places the emphasis on the tier partners to keep itself up to date with progress on the other tiers, as any other approach would be incompatible with rapid time-scales that are demanded.

Despite the fact that, on the face of it, certain aspects of the construction process appear to be a prime candidate for this approach, the biggest obstacles to be overcome by the construction industry in adopting manufacturing-industry-style supply-chain management are as follows:

- Unlike manufacturing, the planning, design and procurement of a building is at present separated from its construction or production.
- The insistence that unlike an airplane or motor car, every building is bespoke, a prototype, and therefore is unsuited to this type of model or, for that matter, any other generic production-sector management technique. This factor manifests itself by:

  ○ geographical separation of sites that causes breaks in the flow of production;
  ○ discontinuous demand; and
  ○ working in the open air, exposed to the elements. Can there be any other manufacturing process, apart from shipbuilding, that does this?

- Reluctance by the design team to accept early input from suppliers and sub-contractors and unease with the blurring of traditional roles and responsibilities.

There is little doubt that the first and third hurdles are the result of the historical baggage outlined in Chapter 1 and that, given time, they can be overcome, whereas the second hurdle does seem to have some validity, despite statements from the proponents of production techniques that buildings are not unique and that commonality even between apparently differing building types is a high as 70% (Ministry of Defence – *Building down Barriers*). Interestingly though, one of the main elements of supply-chain management, Just in Time (JIT) was reported to have started in the Japanese shipbuilding industry in the mid-1960s, the very industry that opponents of JIT in construction quote as an example where, like construction, supply-chain management techniques are inappropriate. Therefore, the point at which any discussion of the suitability of the application of supply-chain management techniques to building has to start is with the acceptance that construction is a manufacturing process, one which can only operate efficiently by means of a managed and integrated supply chain. One fact is undeniable: at present, the majority of clients are required to procure the design of a new building separately from the construction.

Until comparatively recently, international competition, which in manufacturing is a major influencing factor, was relatively sparse in domestic construction of major industrialised countries.

The drivers of supply-chain integration are:

- increased cost competitiveness;
- shorter product life cycles;
- faster product cycles;
- globalisation and customisation of products; and
- higher quality.

One of the principal ways to sustain relationships in the supply chain is through the promotion of collaborative working and transparency.

What identifies a company as having effective and well-managed supply chains in place? The following capabilities should be embedded and evident in the processes of the prime contractor.

At the design stage, the prime contractor should have the ability to:

- carry out an analysis of the project in order to identify supply-chain opportunities;
- prepare strategic sourcing programmes;
- select cross-functional programme management teams; and
- identify and prioritise opportunities and issues.

After this, the needs of the project should be mapped against the capabilities of potential suppliers. The metrics for initial screening involve:

- developing cost models with ring-fenced margins, i.e. profit and offsite overheads;
- supplier selection and development of tier structure;
- defining performance measurement criteria;
- defining client satisfaction; and
- identifying factors that add to performance, value and cost.

Having selected potential suppliers, the prime contractor should be able to move ahead by interrogating the selected supplier's costs paying particular attention to:

- base cost/profit/cost/overheads methodology;
- developing agreements;
- holding implementation workshops;
- developing joint implementation plans; and
- developing systems to measure performance.

Monitoring and progress review systems should be evident. There are two common forms of measurement:

- compliance measurement – did all parties carry out their contractual obligations?; and
- performance measurement – which focuses on measuring performance against targets, time, cost, etc.

There follows a step-by-step approach to applying supply-chain management to sub-contractors:

1. As early as possible:

    - carry out an analysis of the project;
    - identify supply-chain opportunities;
    - prepare a strategic sourcing programme; and
    - identify and prioritise opportunities and issues.

2. Prepare the organisation to engage in supply-chain management:

    - If an organisation has no previous experience in supply-chain collaboration then workshops should be organised in order to ensure that the required levels of enthusiasm are generated. It should be remembered that:

        ○ supply-chain management is not a quick fix and implementation can take time;
        ○ lots of work and investment may be involved in setting up the process; and
        ○ it is important to prepare for resistance and ensure that personnel have ownership of the process.

3. Perform initial supplier screening and meet with suppliers:

    - Company needs should be mapped against the capabilities of potential suppliers. Suppliers should be carefully selected because the company's commitment, in many cases, will be a long-term, intimate business relationship. The metrics for the initial screening involve:

        ○ developing cost models, with ring-fenced margins (i.e. profit and off-site overheads);
        ○ supplier selection and development of tier structure,
        ○ defining performance measurement criteria;
        ○ defining client satisfaction; and
        ○ defining what factors add to performance, value and cost.

4. Supplier selection:

    - In most situations, suppliers are chosen on a combination of the following:

        ○ a track record of demonstrating cost competitiveness and on-time delivery;
        ○ possession of proprietary capabilities;
        ○ demonstrated management capabilities;

  - ○ in-depth quality performance;
  - ○ willingness to develop seamless processes and eliminate waste;
  - ○ compatible cultures;
  - ○ financial strength and profitability; and
  - ○ ability to innovate.

- The initial pool of potential suppliers will be reduced significantly after the initial selection process, and following this, unsuccessful suppliers should be informed why they were unsuccessful. The remaining candidates should then be subjected to in-depth risk assessments to identify strengths, weaknesses and deficiencies.

5. Proceed with selected suppliers:

- Having selected potential suppliers, the process now moves ahead by interrogating the selected supplier's costs paying particular attention to:

  - ○ base cost/profit/cost/overheads methodology;
  - ○ development agreements;
  - ○ holding implementation workshops;
  - ○ developing joint implementation plans; and
  - ○ developing systems to measure performance.

6. Sustain relationships with suppliers:

- One of the principal ways to sustain relationships in the supply chain is through the promotion of collaborative working and transparency. Emphasis should be given to:

  - ○ establishing collaborative relationships with the supply-chain partners; and
  - ○ developing methods to incentivise suppliers to improve their performance, such as pain/gain agreements.

7. Monitor and review progress and improvement:

- There are two common forms of measurement – compliance measurement and performance measurement with regard to aspects such as:

  - ○ time;
  - ○ cost;
  - ○ change;
  - ○ safety;
  - ○ collaborative working;
  - ○ quality; and
  - ○ resources.

- If appropriate, it may be commercially advantageous to expand the relationship into business integration of certain suppliers.

TOOLS AND TECHNIQUES FOR MANAGING THE SUPPLY CHAIN

Supply-chain management is the management, integration and co-ordination of the whole of the supply chain, from, in some cases, the design to the delivery of the complete built asset in order to:

- create a commercially safe environment where the effectiveness of the supply chain as a whole is more important than effectiveness of each individual company;
- deliver projects on time, to budget and to required quality; and
- eliminate anything that does not add value or manage risk.

A contractor's competitive advantage is highly dependent on the integrated management of the supply chain. The supply-chain manager makes use of a growing body of tools, techniques and skills for co-ordinating and optimising key processes, functions and relationships, both with the main contractor and among its suppliers and subcontractors. Supply chains must be flexible and adapt to change in order to survive. A supply-chain manager actively manages the supply chain for the main contractor, a role already well-established in other sectors. Among the attributes that a supply-chain manager can bring to the table is the ability to add value as well as manage risk.

The supply-chain manager's role can be said to operate at two levels:

- The strategic level – this covers:
  - setting up the systems and all that goes with them;
  - relationship management;
  - knowledge sharing;
  - high-level incentivisation;
  - high-level performance measures;
  - high-level targets; and
  - high-level continuous improvement.

- The contract level – this covers:
  - commodity purchases;
  - contract incentivisation;
  - contract-performance measures;
  - contract targets; and
  - contract continuous improvement.

The range of tools and skills commonly used by supply-chain managers includes the following:

- team dynamics;
- capability mapping;

- decision matrices;
- legal frameworks;
- market analysis;
- spend analysis;
- process mapping; and
- negotiation.

Team dynamics

For supply-chain management to be successful, the members of the supply chain have to be serious about implementing supply-chain management and make a commitment to make it work. The team approach must also be extended in order to sustain relationships and deliver continuous improvement.

Capability mapping

Capability mapping is a technique that can be used by supply-chain participants to lay out, in an organised way, all of the criterial functions, processes and capabilities required to design, procure and build the end product. The technique involves mapping key requirements and then superimposing them on one supply-chain map. Maps of requirements can be systematically overlaid with the actual capabilities of each participant to identify gaps, e.g. deficiencies in capabilities and capacity in the supply chain. These potential problem areas then become candidates for more careful data gathering, monitoring and remedial action. Obviously not all functions can be mapped in this way and it is therefore crucial to identify the most significant ones and prioritise the allocation of resources accordingly (see Figure 6.4). The following are examples of capability-mapping parameters:

- demonstrate financial strength commensurate with the risk involved in becoming part of the supply chain;
- ability to provide consistently high-quality products and/or services throughout its own enterprise and throughout its own supply chain;
- compliance with accepted industry standards for quality assessment, quality maintenance and manufacturing performance (e.g. ISO 9000);
- consistent time delivery;
- ability to react to short lead-in times without degrading quality;
- competitive costs;
- ability to design and produce prototypes quickly and accurately; and
- product planning, design and development capabilities.

Decision matrices

Decision matrices can take a variety of formats. They are a method to help sort through various aspects of decision in order to come to the best, most appropriate

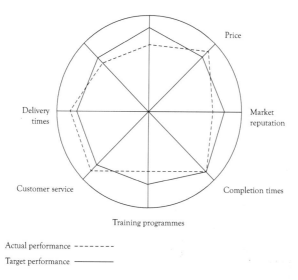

Figure 6.4 Capability mapping

overall choice. They help think about choices, one criterion at a time and then combine these judgements.

Legal frameworks

If tendering is taking place in the public sector then it is likely that the procedures and strategies adopted must currently conform to European Union public-procurement directives.

Market analysis

To manage sub-contractors effectively, the supply-chain manager should identify the competitive forces that shape the construction market and seek to develop a competitive advantage over rival firms. Michael E. Porter's work *Competitive Strategy, Techniques for Analysing Industries and Competitors* (1980) established these forces as:

- the barriers to entry;
- rivalry among existing competitors;
- substitutes;
- the power of the customer; and
- the power of the suppliers.

Spend analysis

Supplier-spend analysis packages are available from a number of specialist providers and are used to reconcile the total spending profile of a contractor. Spend analysis identifies the areas in the supply-management process that can have the biggest impact on profitability. The analysis concentrates on answering the following questions:

- What is spent?
- With whom is it spent?
- How is it spent?

This type of analysis can identify:

- duplicate suppliers;
- where there is too much dependence, and consequently risk, on one sub-contractor or supplier; and
- which sub-contractors or suppliers are dependent on your business and which are critical to the well-being of the main contractor – this sort of information can be useful during negotiations.

Process mapping

Process mapping is complementary to benchmarking and provides a concrete framework within which benchmarks can be set (Figure 6.5). At its most basic level, process-mapping tools define business processes using graphic symbols or objects, with individual process activities depicted as a series of boxes and arrows. Once the processes have been mapped, the next task is to apply appropriate measures to those processes so that points in the process can be identified as high or low risk and the parts of the process that add value can be determined, as well as the identity and extent of the supply-chain drivers.

Negotiation

Of course, not all pricing is achieved through pricing bills of quantities/work packages in direct competition with other firms; there are circumstances where prices are agreed by negotiation. This is true of both the public and private sectors, and some of the main reasons for choosing to negotiate a price are lack of time or works of a highly specialised nature. Two principal methods of progressing a commercial negotiation are:

- direct negotiation: where negotiation is between the client and the contractor/sub-contractor to formulate a contractual and commercial relationship that offers the client a market-competitive deal; or

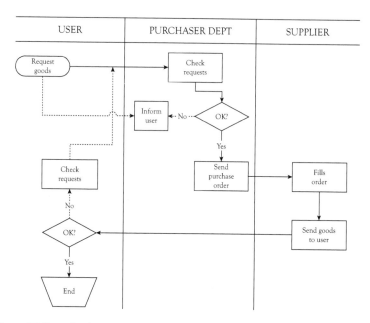

Figure 6.5 Example of process mapping

- indirect negotiation: where negotiation is between a third party acting on behalf of the client and the contractor/sub-contractor to formulate a contractual and commercial relationship that offers the client a market-competitive deal.

Another alternative is competitive negotiation, a further build on the above two approaches that is commonly used in the United States and in the international procurement arena by many of the large corporations having high-procurement spend of the required product/service lines. Competitive negotiation is where a client asks two or more qualified contractors to submit sealed proposals to perform certain work based on a client-prepared proposal package. The client analyses the contractor's proposals, negotiates with two or more of the contractors to arrive at an acceptable scope, terms and compensation for the work, and awards the contract to the contractor whose negotiated proposal represents the highest value to the client per pound spent. The client often uses competitively negotiated contracts when awarding contracts for professional services, reimbursable construction contracts and term contracts.

In using this procurement strategy, it is important to communicate the critical factors, expectations and requirements for consideration by suppliers/contractors, which allow and encourage/solicit creativity and new ideas to find the best way (quality and value) to achieve stated goals. This can be a more flexible approach and differs from developing a competitive bid wherein expectations and requirements are more definitively outlined. The strategy relies on the appropriate negotiating team having strong technical and negotiation skills, as well as developing an effective negotiation plan. Competitive negotiation can be a very effective method for organisations that conduct their business in the international arena and in particular for those organisations having high-procurement spend for their related product/service lines.

Other strategies can include:

- negotiate in parallel with several contractors/sub-contractors (this keeps the competitive threat and your alternatives alive);
- use multiple rounds of negotiations with each contractor;
- negotiate on all commercial terms and conditions – not just price;
- negotiate to achieve the best aspects of all proposals received;
- ask the sub-contractors to suggest ways in which it is possible to lower your internal costs and/or help them lower their costs so they can pass those savings on to the main contractor;
- ask suppliers/sub-contractors what they might do in your situation – you are likely to get very creative approaches to the whole category of sourcing and management;
- negotiate from a 'should cost' basis rather than on market-price discounts; and
- during the negotiations, consistently reassess your information and ask:

    o What is your updated assessment of the bargaining process?
    o Where should you now look for trades?

The key elements of stages of a negotiation are:

- Target the objectives for negotiation:

    o Reduce costs only?
    o Establish a partnership?
    o Long-term agreements?
    o Improve product/service levels?

- Structure for negotiation:

    o What information is needed?
    o Analyses required?
    o Product/material substitution possible?

- Timing for concluding negotiations:

    - Time constraints?
    - Enough time for creativity?

- Contingencies:

    - How to manage sub-contractor/supplier pressure?
    - What happens if sub-contractors/suppliers drop out?
    - How to deal with exceptions?
    - Will late offers be accepted?

7

Applied estimating

In addition to pricing bills of quantities or work packages, there are other situations were estimating/pricing skills are required to be applied. Among these are:

- pro-rata pricing;
- approximate quantities;
- builder's quantities; and
- BIM and estimating software packages.

PRO-RATA PRICING

The variation account

Section 5 of the JCT (16) deals with variations to the contract. After a contract has been signed, it cannot be changed or varied by the parties. However, given the nature of the construction process, with all of its inherent risk and uncertainty, most standard forms of construction contract include the provision for variations or alterations to the works.

The JCT (16) form of contract clause 5.1 defines the term variation as follows:

- the alteration or modification of the design, quality or quantity of the works including:

  o the addition, omission or substitution of any work;
  o the alteration of the kind or standard of any of the materials or goods to be used in the works; and
  o the removal from the site of any work executed or materials or goods that are not in accordance with the contract.

In addition, the contract allows for variations or alterations to obligations and restrictions imposed on the contractor such as access to the site, working hours, etc.

The variation account will probably be the largest, in terms of documentation, in a final account. The number and nature of change orders or variations issued during the contract period will vary considerably according to the type of contract. For example, contracts with little or no pre-contract planning/documentation or refurbishment contracts can expect to generate a substantial number of variations whereas comparatively uncomplicated projects based on fully detailed documentation may generate very few.

The procedure for issuing a variation order using JCT (16) is as follows:

- All variations must be in writing and issued by the architect/contract administrator. If the contractor objects to carrying out the variation then he/she has seven days to object. Any objection must be in writing. If the contractor fails to comply with the instruction within seven days then the architect /contract administrator has the right to get another contractor to complete the variation and contra-charge the contractor.
- In addition to architect's instructions, variations may also be given to the contractor in the form of an oral instruction or a site instruction, given by the clerk of works. Both of these forms of variation have to be confirmed in writing by the architect/contract administrator in order to become an official variation.
- Some variations will be issued by way of a drawing together with a covering architect's instruction, and there may be occasions where several revisions of the same drawings may be issued. In circumstances such as these, it is important that all drawings are carefully logged in a drawing register and only the latest revision is used to measure and assess the variations/alterations. Note that to be valid, drawings must be signed by the architect or accompanied with a covering letter – see *Myers* v. *Sarl* (1860).
- The architect may not omit work that has been measured in the bills of quantities and priced by the contractor, in order to give it to others to carry out.
- Any errors or omissions that come to light during the contract period are, in the case of JCT (16) clause 2.14/15, dealt with as a variation order.

The valuing of variations is generally divided into two operations:

- measurement and pricing of variations; and
- pro-rata pricing.

Measurement and pricing of variations

The JCT (16) clause 5.4 gives the contractor the right to be present when variations are measured. In practice, measurement of variations can take place: on site; by the employer's quantity surveyor who then passes it to the contractor for checking and agreement; or by the contractor as part of the interim valuation process, which has to be checked and agreed by the employer's quantity surveyor. Once agreed, variations are included in the final account and included in interim certificates.

JCT (16) clause 5.6 sets out the rules for valuing variations, which are a set of procedures in descending order of preference, namely:

- Where the additional works are of similar character to items in the original contract bills, then the bill rates are used.
- Where the additional works are different in as much that they are carried out under different conditions or there is a significant change in the quantities, then the bill rates are used in order to prepare a fair price, usually by the build up of pro-rata prices (see next section).
- If neither of the above two approaches is appropriate, then work shall be valued at fair rates and prices.
- Finally, under certain circumstances defined in clause 5.7, dayworks may be used as the basis for valuation (see Chapter 5).

No allowances for any effect on the regular progress of the work caused by the issue of a variation order can be built into the pricing. Any such claims have to be the subject of a claim for loss and expense.

From the quantity surveyor's view point, it is best to measure and value variations as quickly as practical so that increases/decreases can be built into financial statements and, finally, accounts. From a logistical point of view, it is good practice to group similar items together; for example, work to the substructure or drainage being measured in one omnibus item.

Pro-rata pricing

With respect to the valuation of variation, the JCT (16) refers to taking into account similarity, character, conditions and quantity of work involved, and this can be problematic at times.

One of the provisions in the JCT (16) form is to price variations, where no direct price is available, at rates that are based on the bill rates; this is known as pro-rata pricing and, as with approximate quantities, there is a number of approaches.

As explained in Chapter 3, bill rates are an amalgam of materials, labour, plant, profit and overheads, therefore when attempting a pro-rata calculation a known built-up rate is broken down into its constituent parts in order that it can be adjusted. If too many of the elements that constitute a rate have to be adjusted then it is better to build up a new rate from basic principle as described in Chapter 4.

One of the advantages of having a bill of quantities is that the degree of detail contained in the document can be used as a basis for the valuing of variations during the post-contract stage of a project. As discussed in the previous pages, bill rates are composed from the following:

- labour;
- materials;

- plant; and
- profit and overheads.

The technique of pro-rata pricing involves disassembling a bill rate and substituting new data in order calculate a new rate that can be used for pricing variations.

Example

The client's quantity surveyor is meeting with the contractor to price a variation that has been measured previously. It is not possible to directly use the existing bill rates to price an item, therefore a pro-rata price is built up. As the contractor is present, it may be possible to refer back to the original price build-up and this will make the process easier. There are occasions, however, when the quantity surveyor works alone and therefore appropriate assumptions have to be made. The approach is to analyse the original bill rate as follows:

- deduct profit and overheads;
- analyse materials, labour and plant costs;
- adapt costs to suit new item; and
- add back profit and overheads.

Bill of quantities item

38 mm thick cement and sand (1:3) screed exceeding 600 mm wide level and to falls only not exceeding 15° from horizontal to concrete with steel trowel finish £27.38/m².

Variation account item

50 mm thick cement and sand (1:3) screed exceeding 600 mm wide level and to falls only not exceeding 15° from horizontal to concrete with steel trowel finish. Cost per m².

Analysis of bill rate	£	£
38 mm thick cement and sand screed	£27.38	
Less profit and overheads 15%	£ 3.57	
Net cost	£23.81	23.81
Deduct materials		
1 m³ cement = 1,400 kg cement @ £190.00 per tonne	266.00	
3 m³ sand = 4,800 kg @ £15.00 per tonne	72.00	
	338.00	
Add shrinkage 25%	84.50	
	422.50	
Add waste 5%	21.13	
Cost per 4 m³	£433.63	
÷ 4 – cost per m³	£108.41	

Mixing

Assume 100 litre mechanical fed mixer @ £24.00	<u>6.00</u>	
per hour; output 4 m³ per hour – cost per m³		
Cost per m³	£114.41	
Cost per m² – 38 mm thick		<u>£4.48</u>
Cost of labour for 38 mm thick screed		£19.33

Materials

Cost per m³ as before	<u>114.41</u>	
Cost per m² – 50 mm thick		<u>£5.72</u>
		£25.05

Labour

Labour costs for a 50 mm thick screed will be	<u>£3.87</u>
approximately 20% more expensive to place a	
38 mm screed £19.33 × 0.20	
	£28.92
Add profit and overheads 15%	<u>£4.34</u>
Cost per m² for 50 mm thick cement and sand (1:3) screed	**<u>£33.26</u>**

Example

Bill item

300 × 125 mm precast concrete (Mix A) weathered and twice-throated coping bedded in gauge mortar (1:1:6) £35.00 per m

Variation account item

225 × 150 mm precast concrete (Mix A) weathered and twice-throated coping bedded in gauge mortar (1:1:6). Cost per m.

For a straightforward item such as this, the only real variation will be the cost of the smaller sized coping as the labour involved will be the same as the bill item.

Calculation

Bill item	£35.00
Less profit and overheads 15%	<u>£4.57</u>
Net cost, labour and materials	£30.43
Deduct cost of 300 × 150 mm coping	<u>£19.25</u>
Add cost labour	£11.18
Add 225 × 150 mm coping	<u>£15.75</u>
	£26.93
Add profit and overheads 15%	<u>£5.54</u>
	<u>£32.47</u>

APPROXIMATE QUANTITIES

This is regarded as the most reliable and accurate method of estimating, provided that there is sufficient information to work on. Depending on the experience of the surveyor, measurement can be carried out fairly quickly using composite rates to save time. The rules of measurement are simple, although, it must be said, not standardised, and tend to vary slightly from one surveyor to another:

- One approach involves grouping together items relating to a sequence of operations and relating them to a common unit of measurement, unlike the measurement for a bill of quantities, where items are measured separately.
- Composite rates are then built up from the data available in the office for that sequence of operations.
- All measurements are taken as gross overall, with the exception of the very large openings.
- Initially, the composite rates require time to build up but once calculated they may be used on a variety of estimating needs.
- Reasonably priced software packages are now available.

The following is an example of a composite for substructure.

Example

Excavate trench width exceeding 0.30 m maximum depth not exceeding 1.00 m; earthwork support, filling to excavations, disposal of excavated material off site, 450 × 600 mm in situ reinforced concrete (30N/mm2 aggregate) edge beam, 275 mm cavity brickwork to 150 mm above ground level, bitumen-based damp-proof course, facings externally. Cost per m.

Normally, if measured in accordance with NRM2, these items would be measured separately; however, when using approximate quantities a composite rate is calculated that includes a mix of units of measurement and is applied to a linear metre of trench. Note that some of the items in the above description are deemed to be included – that's to say they do not have to be specifically measured in accordance with NRM2 – nevertheless, these items, for example earthwork support, may have to be included to comply with health and safety and therefore allowed for in the price build-up.

Build-up of approximate quantities rate

			£
Excavation to receive edge beam	1.00		
	0.45		
	<u>1.00</u>	0.45 m³ @ £15.34 m³	6.90

Remove surplus spoil	1.00 0.17 <u>0.70</u> =	0.12 m³ @ £5.73 m³	0.69
Backfilling	1.00 0.37 <u>1.00</u> =	0.37 m³ @ £7.00 m³	2.59
Earthwork support	1.00 <u>1.00</u> 1.00 × 2 =	2.00 m² @ £8.91 m²	17.82
Compacting excavation	1.00 <u>0.45</u> =	0.45 m² @ £3.00 m²	1.35
25N/mm² concrete edge beam	1.00 0.45 <u>0.30</u> =	0.14 m³ @ £176.46 m³	24.70
Bar reinforcement		5 kg @ £2.00/kg	10.00
Formwork	1.00 <u>0.30</u> × 2 =	0.60 m² @ £38.48 m²	23.09
275 mm cavity wall	1.00 <u>0.85</u> =	0.45 m² @ £176.97 m²	79.64
Damp-proof course	<u>1.00</u> × 2 =	2 m @ £ 4.78 m	<u>9.56</u>
Cost per m			£176.34

Therefore, to estimate the cost of the total substructure, calculate the total length of trench excavation and concrete, etc. and multiply by £176.34. To complete the substructure estimate, calculate the cost of the oversite slab per square metre.

BUILDER'S QUANTITIES

Builder's quantities are quantities measured and described from the builder's view point, rather than in accordance with a set of prescribed rules, such as SMM7 or NRM2. There can be quite a big difference between the builder's quantities and approximate quantities. For example, earthwork support may be required to be included, whereas in practice this may be omitted from the site operation with the sides of the excavation being battered back instead. A more pragmatic approach is reflected in the measurement and pricing of builder's quantities.

Running and maintenance costs

NRM3: Order of Cost Estimating and Cost Planning for Building Maintenance Works was published in February 2014, and became effective on 1 January 2015.

NRM3 provides guidance on the quantification and description of maintenance works for the purpose of preparing initial order of cost estimates during the preparation stages of a building project, elemental cost plans during the design development stages, and detailed asset-specific cost plans during the procurement of construction and the post-construction or in-use phases of a building project or facility.

The rules follow the same framework and premise as NRM1: *Order of Cost Estimating and Cost Planning for Capital Building Works*. Consequently, they give direction on how to quantify and measure other items associated with maintenance works, but that are not reflected in the measurable maintenance work items – i.e. maintenance contractor's management and administration charges, overheads and profit, other maintenance-related costs, consultants' fees, and risks in connection with maintenance works.

Unlike capital building works projects, maintenance works are required to be carried out from the day a building or asset is put to use until the end of its life. Accordingly, while the costs of a capital building works project are usually incurred by the building owner/developer over a relatively short term, costs in connection with maintenance works are incurred throughout the life of the building – over the short, medium and long terms. Consequently, the rules provide guidance on the measurement and calculation of the time value of money, guidance on using the measured data to inform life-cycle cost plans and forward maintenance plans, as well as VAT and taxation.

NRM3, together with NRM1, presents the basis of life-cycle cost management of capital building works and maintenance works – enabling more effective and accurate cost advice to be given to clients and other project team members, as well as facilitating better cost control.

NRM3 has been written to provide a standard set of measurement rules that are understandable by all those involved in budgeting for, cost managing and procuring construction and maintenance works on discrete buildings, building portfolios, establishments and/or estates, including the employer, thereby aiding communication between the project team and the employer and parties associated with the delivery of future maintenance works. In addition, NRM3 should assist the quantity surveyor/cost manager, as well as the facilities manager, in providing effective and accurate cost advice to the employer and other project stakeholders throughout the life-cycle cost-management process.

The document provides rules of measurement for the preparation of order of cost estimates and elemental cost plans during construction procurement and *asset-specific cost plans* for maintenance works undertaken post practical completion (i.e. during the 'use' of the building). Direction on how to describe and deal with costs and allowances forming part of the cost of maintaining a building or constructed asset or its parts, but that are not reflected in the measurable maintenance work items, is also provided.

The rules also provide a cost-management framework that can be used to develop labour-resource plans for the annualised maintenance works and for predicting the intermittent or periodic forecast of life-cycle major repairs and replacement (renewal) works for the defined period of analysis.

NRM3 is divided into five parts with supporting appendices.

Part 1

- Places order of cost estimating and cost planning in context with the RIBA Plan of Work and the OGC Gateway process.
- Defines the purpose, use and structure of the rules.
- Clarifies how maintenance relates to other aspects of life-cycle costing.
- Details cost categories and definitions for the constituents' maintenance works.
- Provides preparation rules for defining the brief and project particular requirements.
- Provides guidance on dealing with projects comprising multiple buildings or facilities.
- Explains the symbols, abbreviations and definitions used in the rules.

Part 2

- Describes the measurement rules for *order of cost estimating* for maintenance works.
- Explains the purpose and methods used and information required to do an order of cost estimate.
- Defines its key constituents.
- Explains how to prepare an order of cost estimate using the floor area method and functional unit method.
- Provides measurement rules for annualised maintenance and periodic life-cycle repairs and replacements, during the initial phases of a construction project.
- Provides the measurement rules for dealing with the maintenance contractor's management and administration, overheads and profit, other project-specific costs/consultant fees, employer-definable maintenance-related costs, risks and time value of money – plus VAT taxation and other considerations.
- Provides guidance on rules on cost reporting of order of cost estimates, including the analysis and benchmarking of maintenance works cost.

Part 3

- Describes the measurement for *cost planning* for maintenance works.
- Explains the methods used for *elemental cost plans* and *asset-specific costing*.
- Explains the constituents of an elemental cost plan.
- Defines the information required to enable preparation of elemental cost planning during construction procurement and for the *asset-specific costing* during in use phases of buildings or constructed facility.
- Defines the method of dealing with the maintenance contractor's management and administration costs, overheads and profit, project/consultant fees, other relevant maintenance/project costs, risks, time value of money and taxation.

Part 4

- Describes the measurement for *asset-specific cost planning* for maintenance works.
- Explains the methods used for renewal cost plans generated from capital building cost plans and from asset-survey-based cost information.
- Explains the measurement rules for maintenance cost planning based on asset-specific planned maintenance tasks and provision for unscheduled works.
- Gives cost-guidance rules for asset-data collection, physical condition and remaining life surveys, and other forms of assessment.

Part 5

- Comprises the tabulated rules of measurement and quantification of asset-specific costing for annualised maintenance and periodic life-cycle repairs and replacement works.
- Provides UK standardised cost structures for maintainance and renewal works.
- Describes methods of codification of cost planning of annualised maintenance and service-life planning of renewal works.
- Provides level codes and naming conventions.
- Describes methods of coding for asset surveys and condition/remaining-life assessments.
- Highlights the need to apply specific maintenance strategies and to undertake planned inspections of buildings and services to identify and quantify the repairs and replacement works.

BIM AND ESTIMATING SOFTWARE PACKAGES

Cost estimating in BIM

One of the major developments in project-data management to emerge during the past ten years is BIM. In 2011, the government announced that all suppliers that wished to bid for public-sector building contracts must use BIM tools and techniques from 2016, making its implementation commercially critical for companies wishing to apply for high-value public projects in the future. As the latest 2018 NBS survey shows, the uptake of BIM by the construction industry to date has not matched the initial hype, with only 18% of respondents indicated they used BIM on every project, which is a slight decrease on the previous year.

BIM is the process of bringing together and sharing information in a digital format among all those involved in a construction project, including architects, engineers and builders. By making information far more accessible and available to the client and end user to support through-life asset management, BIM is claimed to be a path to greater productivity, risk management, improved margins and sustainability. BIM envisages virtual construction of a facility prior to its actual physical construction in order to reduce uncertainty, improve safety, work out problems, and simulate and analyse potential impacts. Sub-contractors from every trade can input critical information into the model before beginning construction, with opportunities to pre-fabricate or pre-assemble some systems off-site. Waste can be minimised on site and components delivered on a JIT basis rather than stockpiled on site.

Estimators know there is more to cost estimating in BIM than simple automation of estimating from objects to spreadsheets. Estimators also understand the challenges and obstacles beyond the technology that must be overcome if cost estimating is to become a viable dimension of BIM. For purposes of this section, there are two basic aspects of BIM relevant to estimating:

- BIM is intended to improve industry efficiency and productivity with accurate and complete information; and
- BIM should support the entire life cycle of a facility, and therefore information contained in the model should facilitate the work of all stakeholders.

The industry associations of professional estimators are working together with the Building Smart Alliance (BSA) to define and develop processes in BIM that bridge the gap between traditional and BIM processes. These associations include the Association for the Advancement of Cost Engineering, the American Society of Professional Estimators, and the RICS. This group responds to the increasing demand for cost estimating in BIM. The following is a brief description of the traditional estimating process and some changes required in the process to produce valid and accurate estimates using BIM.

Traditional and BIM: similarities and differences for estimators

One convention employed by estimators in the traditional process is in identifying the expected accuracy range of an estimate based on the level of project detail. In the traditional process, the project drawings and specifications were the primary means by which this was determined, and, as such, there was a direct correlation between the project's level of detail and the expected accuracy of an estimate. It is reasonable to expect that a similar convention exists in BIM, and that as a BIM model contains more project detail, it also impacts the potential accuracy of an estimate. The difference in BIM, though, is in how a designer creates the objects for drawings and specifications now has an impact on the estimate.

The method or sequence by which a designer created plans and specifications in the traditional formats did not impact the estimate because the information relevant to an estimate was an overlay by the estimator and external to the graphical representation. In the traditional process, the estimator managed the information from these documents and extracted, organised and used the information as best suited to accomplish the task of estimating. However, with

BIM the point of organising information shifts as more of it begins in the design-model phase. To date, there is no industry standard to bridge the gap between the design model and the estimator. Consequently, this has an effect on estimator's confidence about the information in a BIM model and its fitness for purpose relative to estimating.

This is a challenge as estimators are now faced with the accuracy level of estimates dependent on the validity and reliability of information in the model. Model objects are rich with the information estimators need to create a cost estimate, and if this information is to be used by estimators, then there is a point where the estimator's process should filter into the information management during design.

The development of a BIM model includes the graphical representation of data-rich objects. The primary purpose of the design model is to convey design intent. However, each of the objects inserted are available now for future extraction by other stakeholders. The difference in BIM is that from an estimator's perspective, the development of a model is about the information associated with the objects and the input process for this information. This aspect of BIM is a significant shift in paradigm from which the estimator previously worked. It is crucial that the estimator has confidence that the information is a valid representation of the object beyond the model to physical reality. This is new in the world of estimating and is challenging estimators as they work within this new paradigm.

The estimator's responsibility, as the sole individual responsible for organising information in a way useful for an estimate, is different in BIM compared to the traditional process. The traditional responsibility of the estimator managing the information for an estimate is distributed now to earlier phases in a project simply by the nature of BIM. If estimators are not involved at earlier phases, it leads to redundancy and inefficient work processes. Currently, estimators working in BIM often invest valuable time in preparing or revising the model to facilitate its use as a tool for legitimate output in 3D/5D. Not only is rework inefficient, it also increases the potential opportunity for errors. More importantly, this activity adds no value to the project and mitigates the efficiencies intended to be associated with BIM. Reworking BIM data by the estimator goes against one of the basic principles of BIM, namely to increase efficiency and productivity.

As estimating in BIM continues to develop, it is important to keep in mind that traditional cost estimating goes beyond material quantity take-offs and pricing. It includes the modelling of project construction with conditions and constraints that impact the construction process. As such, cost estimators build the facility using quantity take-offs of specified materials, then add professional knowledge of means and methods, sequencing and phasing, conditions, and constraints. This is the cost-estimating process, and inputs from other stakeholders are embedded in this process. The previous methods used for developing plans to convey design did not impact the estimator. The estimator's ability to complete an accurate estimate using BIM is a challenge at this time. A major obstacle is the lack of a standard that establishes how a BIM model is created so that it contains valid and reliable information to meet the needs of all stakeholders across the life cycle of a facility.

By definition, a BIM model is an intelligent model, and it is logical to expect that the information within the model goes beyond the needs of the creator who inserted the information to facilitate other stakeholders' tasks and work processes. A BIM model should be developed for the life cycle of a facility and with information input from multiple stakeholders for extraction and use by multiple stakeholders. It is important to think beyond the usefulness of information for one's own use to how the information will be used by others.

Estimating has always relied on the inputs from the design process, and this remains unchanged. The difference in BIM is that the method and organisation of the inputs by the design team have an impact beyond the design process. All inputs are rich with information and available for other stakeholders, and as such the co-ordination of information at all points in a BIM model is important. As 3D/5D gains momentum and more projects require estimating in BIM, professional estimators are working to define a new process and capitalise on the opportunities available for improving cost estimating in BIM.

Estimating software

Producing a construction bid, whether it's for a small residential project or a big civil project, can be a lengthy and time-consuming operation. Construction estimating software will not only simplify the process, but also it can help increase productivity and limit the number of errors in bids, as well as tracking and comparing the estimated costs with actual costs.

Using construction-estimating software can give a competitive advantage over other bidders. It allows contractors to manage job costs as well as sub-contractor bids, all while monitoring profits. Every aspect of construction, plant, labour and materials will be accounted for and included in the bid, thereby reducing the impact of unforeseen costs.

Some of the more popular BIM-based estimating systems that are available are:

- Autodesk Quantity Takeoff cost estimating tool 2018; and
- CostX.

Estimating systems are generally add-on software packages to systems offering such functions as taking-off quantities and visualisation and based on Revit 2019.

BIM-based estimating systems offer the following:

- a speeded-up process by automating measurement;
- links from the costs and quantities to a digital model;
- greater understanding of what is being priced and what the client is getting for their investment through visualisation;

- the ability to highlight objects or building components within the model;
- changes in the BIM model can be related to the estimate; and
- easy analysis.

However, on the negative side, basic data still have to be calculated and imported into the system.

What to look for

Estimating software must be able to prepare and monitor outgoing bids efficiently and accurately. The best construction-estimating software will also provide the ability to complete a number of other tasks that constitute the daily business functions of a busy contractor. Here are a few of the criteria we used to evaluate each one of these estimating applications:

FEATURES

Estimating software should feature tools that will make estimating easier. This includes the ability to track job costs and bids, as well as to perform general office-management functions such as document preparation, purchase-order processing and archiving previous projects. It should also perform some accounting tasks and create invoices. Good estimating software gives the flexibility needed to create bids for a variety of different projects. It should also accept customised formulas and calculations, provide reports, import national cost indices, compare sub-contractor and vendor quotes, and track jobs by divisions and sub-divisions.

EASE OF USE

Estimating software manages many variables and, therefore, can be quite complex, but using the software should be as easy as possible. The estimating application should be well organised and presented in a logical manner so that users get the most out of this software in a short amount of time.

INTEGRATION

High-quality software can seamlessly integrate with other third-party applications. Integration with accounting software such as QuickBooks, CAD software, Microsoft Project and more will enable a contractor or sub-contractor to save time and money by consolidating resources. Additionally, compatibility with mobile devices such as iPads and smartphones will allow the updating of project information on the go.

HELP AND SUPPORT

Since this type of software involves many aspects, the manufacturer should provide tutorials, training classes or online demonstrations to help users learn how to utilise all of the features in a productive manner. Customer support should be readily available in a variety of means, and the support team should respond to enquiries quickly and thoroughly. Investing in construction-management software will help increase speed and accuracy in creating and presenting bids. While no estimating-software application can guarantee winning every bid, using estimating software can help create the best bid possible.

POST-TENDER ESTIMATING

A post-tender estimate is mid-way between tender reporting and cost reporting. It is prepared during RIBA Plan of Work stage, H (Tender Action), or OGC Gateway 3C (Investment Decision), after the construction tenders have been received and evaluated. The purpose of a post-tender estimate is to forecast the cost of the project using a similar model to a cost report but in the case of the post-tender estimate, the scope of works should take into account as many cost centres as possible.

In some cases, it may be incorporated into the tender report or it can become the first cost report. It is, however, an entity in its own right. It is pre-contract but post-tender. The aim of this estimate is to confirm the funding level required by the employer to complete the building project.

The post-tender estimate is based on:

- the results of any post-tender negotiations, including the resolution of any tender qualifications and tender-price adjustments;
- any actual known construction costs;
- any residual risks;
- site issues that may have been uncovered through further ground and site investigations;
- client variations and instructions that could not have been incorporated at tender stage; and
- cost updates of project and design-team fees, as well as other development and project costs, where they form part of the costs being managed by the cost manager.

When reporting the outcome of the tendering process to the employer, the quantity surveyor/estimator should include a summary of the post-tender estimate(s). The post-tender estimate should be reasonably accurate because the uncertainty of market conditions has been removed.

Post-tender estimates are used as the control estimate during construction.

Appendix A

Indicative costs per m² for a range building types

Building type	£/m²
Residential	
Housing – private	
Bungalows	826–1,545
Semi-detached housing	890–1,390
Detached housing	970–2,020
Flats	700–1,400
Educational	
Nursery school	800–2,100
Primary school	1,280–2,150
Secondary school	1,034–1,925
University	1,570–2,500
Adult education	1,190–2,100
Public library	1,800–2,400
College	1,100–2,000
Laboratory	1,600–2,900
Commercial/administration	
New-build offices	
Non-air-conditioned	930–1,700
Air-conditioned	1,200–1,900
Prisons	2,300–2,800
Industrial	
Factories	630–890
Warehouses	430–880

Building type	£/m^2
Health	
General hospitals	2,100–3,200
Day hospitals	1,900–2,400
Day centres	1,500–2,230
Retail	
Warehousing	
1,000–5,000 m^2	430–850
5,000–15,000 m^2	350–750
Religious	
Churches	950–2,450
Mixed use (offices, shops, flats)	
Three–five storey	1,160–1,600
Six storey +	1,300–2,000
Leisure	
Pub	1,300–1,700
Community centre	
Timber framed	2,300–2,700
General purpose sports hall	1,100–1,700
Cinema	1,580–1,700
Recreational	
Canteen	1,180–2,500
Theatre	800–1,300
Squash courts	1,100–1,700

The above figures are based on second quarter 2018 prices and should be adjusted accordingly with the appropriate indices.

Appendix B

Useful rules and conventions

EARTHWORKS

Increases in bulk of excavated materials:

Rock	50%
Chalk	33%
Clay	25%
Sand/gravel	10%

For excavation other than medium clay, the following multipliers should be applied to labour constants:

Loose sand	0.75
Stiff clay or rock	1.50
Soft rock	3.00

HARDCORE FILLING

If bought by volume, an average of 20% should be added to the material to cover consolidation and packing. If bought by weight, the following is the approximate amount required per cubic metre of compacted material:

Brick ballast	1,800 kg
Stone ballast	2,400 kg

CONCRETE WORK

Material	Tonnes/m³
Cement in bags	1.28
Cement in bulk	1.28–1.44
Sharp sand (fine aggregate)	1.60
Gravel (coarse aggregate)	1.40

Mixer outputs in m³ of concrete per hour

Hand loaded:

100 litre	150 litre	175 litre	200 litre
1.2	1.8	2.1	2.4

Mechanically fed:

200 litre	300 litre	400 litre	500 litre
4.0	7.2	9.6	12.0

MASONRY

Bricks per m² based on 215 × 102.5 × 65 mm bricks

Half brick wall:	
Stretcher bond	60
One-brick wall:	
English bond	120
English bond	90
Flemish bond	80

Mortar per m²

Brickwork

	55 mm brick	65 mm brick	75 mm brick
Per 102.5 mm thickness of wall	0.035	0.03	0.028
Per vertical joint (perpend)	0.01	0.01	0.01

Blockwork

440 × 215 × 100 mm blocks – amount of mortar per m² of blockwork – 0.01 m³

CARPENTRY

Item	Hours per m (carpenter)
Wall plates	0.10
Floor joists	0.15
Partitions	0.30
Ceiling joists	0.15
Flat-roof joist	0.17
Strutting	0.30

ROOFING

To calculate the number of centre nail slates size 450 mm × 300 mm, laid with 100 mm lap:

$$\frac{\text{length of slate} - \text{lap}}{2} \times \text{width of slate}$$

$$\frac{0.450 - 0.100}{2} \times 0.300 = 0.0525$$

$$\frac{1 \text{ m}^2}{0.0525} = 19.05 \text{ say } 19 \text{ slates per m}^2$$

The number of head-nailed slates per square metre:

$$\frac{\text{length of slate} - \text{lap} - 25 \text{ mm}}{2} \times \text{width of slate}$$

Clay concrete tiles

	Lap	Gauge	No./m²	Battens m/m²
267 × 165 mm	65	100	60	10
	65	98	64	10.5
	65	90	66	11.3

	Lap	Gauge	No./m²	Battens m/m²
387 × 230 mm	75	300	16	3.2
	100	280	18	3.5
420 × 330 mm	75	340	10	2.9
	100	320	11	3.2

Labour constants and materials for plain tiling per m²

Lap	Tile nails (kg)	Batten nails (kg)	Tiler and labourer
65	0.09	0.11	0.67
75	0.10	0.12	0.73
90	0.11	0.13	0.79
100	0.12	0.14	0.83

ASPHALT

Asphalt is manufactured in 25 kg blocks and then heated to melting point and applied on site. The following is the approximate covering capacity of 1,000 kg of asphalt;

First 12 mm thickness	35 m²
Additional 3 mm thickness	150 m²

PLASTERING

Approximate coverage per tonne	m²
11 mm Browning plaster	135–155
11 mm toughcoat plaster	135–150
11 mm hardwall plaster	115–130
2 mm board finishing plaster	410–430
2 mm finishing plaster	410–430
13 mm universal one coat	85–95

DECORATION

A roll of UK wallpaper is 0.53 m wide and 10.05 m long, which equals 5.33 m².

DRAINAGE

The following is a guide to the width of a drain trench based on pipes up to 200 mm diameter:

Depth of trench	Width
Average depth up to 1 m	500 mm
Average depth 1 m to 3 m	750 mm
Average depth over 3 m	1000 mm

Further reading

1 Introduction

Cartlidge D. (2018). *New Aspects of Quantity Surveying Practice – Fourth Edition,* Routledge, Oxford

Cartlidge D, (2017). *Quantity Surveyor's Pocket Book – Third Edition,* Routledge, Oxford

Dalziel B. Ostime N (2013) *Architect's Job Book.* RIBA Publications

Office of National Statistics (2018) *Annual Construction Statistics.* HMSO

The RICS (2012), *RICS New Rules of Measurement, Order of cost estimating and elemental cost planning, 2 ndEdition,* RICS Books

The RICS (2012), *RICS New Rules of Measurement, Bill of Quantities for Works Procurement, 1st Edition*, RICS Books

The Institution of Civil Engineers (2012) *Civil Engineering Standard Method of Measurement – Fourth Edition,* ICE Publishing

NBS (2018), *NBS National Construction Contracts and Law 2018*

NBS (2018), *NBS BIM Report 2018*

2 Early cost advice

The RICS (2007) *The RICS Code of Measurement Practice 6th Edition,* RICS Books

RICS professional statement (2018); *RICS property measurement 2nd edition*

The RICS (2013) *New Rules of Measurement: Order of cost estimating and cost planning for building maintenance works,* RICS Books

3 Resources

CIOB (2018) *New Code of Estimating Practice.* Wiley-Blackwell

The RICS (2012), *RICS New Rules of Measurement, Bill of Quantities for Works Procurement, 1st Edition*, RICS Books

BATJIC (2018) *National Joint Council for the Construction Industry's Working Rule Agreement,* Building and Allied Trades Joint Industrial Council

4 Unit rate estimating

Cartlidge D. (2004) *Procurement of Built Assets,* Butterworth Heinemann, Oxford

Hughes W. Hillebrandt P.,Greenwood D. (2006) Procurement in the Construction Industry. Taylor and Francis

CIOB (2018) *New Code of Estimating Practice.* Wiley-Blackwell

5 Tender settlement

Brook M (2016) *Estimating and Tendering for Construction Work 5th Ed.* Routledge

CIOB (2018) *New Code of Estimating Practice.* Wiley-Blackwell

Finch R (2011) *NBS Guide to Tendering for construction projects,* RIBA Publishing

The RICS (2007) *RICS/BEC definition of Prime Cost of Daywork 3rd Edition,* RICS Publishing

The RICS (2010) *Schedule of Basic Plant Charges for Use in Connection with Daywork Under a Building Contract* RICS Publishing

The RICS (2011) Definition of Prime Cost of Daywork Carried Out Under An Electrical Contract - Third Edition, RICS Publishing

The RICS (2012) *Definition of Prime Cost of Daywork Carried out under a Heating, Ventilating, Air Conditioning, Refrigeration, Pipework and/or Domestic Engineering Contract – Second Edition,* RICS Publishing

The RICS (2007) *Definition of Prime Cost of Daywork Carried Out Under A Plumbing Contract – First Edition*, RICS Publishing

6 The supply chain

Chappell D. (2006) *Construction Contracts Questions and Answers.* Taylor and Francis

Hacket M.,Robinson I.,Statham G. (2016) Aqua Group *Guide to Procurement Tendering and Contract Administration. Blackwell Publishing 2nd Ed.*

RICS Quantity and Construction Professional Group Board (2012) *Black Book – Various Chapters*, RICS Publishing

Michael E. Porter's (1980) *Competitive Strategy, Techniques for Analysing Industries and Competitors,* Simon and Schuster.

7 Applied estimating

Dennis M. Gier, MS, PE (2007) *What Impact Does Using Building Information Modeling Have on Teaching Estimating to Construction Management Students?* California State University

Crotty R (2011) *The Impact of Building Information Modeling Transforming Construction,* Routledge

NBS (2018), *NBS BIM Report 2018*

Song Wu, Gerard Wood, Kanchana Ginige, Siaw Wee Jong (2014), *A TECHNICAL REVIEW OF BIM BASED COST ESTIMATING IN UK QUANTITY SURVEYING PRACTICE, STANDARDS AND TOOLS,* Journal of Information Technology in Construction

Index